AF835037

Learning Energy Efficiency Networks

Learning Energy Efficiency Networks

Applications and Challenges in Developing Countries

edited by
Claudio Carpio
Marina Yesica Recalde

JENNY STANFORD
PUBLISHING

Published by

Jenny Stanford Publishing Pte. Ltd.
101 Thomson Road
#06-01, United Square
Singapore 307591

Email: editorial@jennystanford.com
Web: www.jennystanford.com

British Library Cataloguing-in-Publication Data
A catalogue record for this book is available from the British Library.

Learning Energy Efficiency Networks: Applications and Challenges in Latin American Countries

Copyright © 2026 by Jenny Stanford Publishing Pte. Ltd.
All rights reserved. This book, or parts thereof, may not be reproduced in any form or by any means, electronic or mechanical, including photocopying, recording or any information storage and retrieval system now known or to be invented, without written permission from the publisher.

For photocopying of material in this volume, please pay a copying fee through the Copyright Clearance Center, Inc., 222 Rosewood Drive, Danvers, MA 01923, USA. In this case permission to photocopy is not required from the publisher.

ISBN 978-981-4968-82-9 (Hardcover)
ISBN 978-1-003-77468-6 (eBook)
DOI 10.1201/9781003774686

List of Contributors

Angel Alfredo Núñez Barreto
National Office for the Control of the Rational Use of Energy (ONURE), Matanzas, Cuba
mt.lopez@nauta.cu, dtorure@electmtz.une.cu

Daniela Méndez Bellamy
Senior international consultant
dmendez.bellamy@gmail.com

Jairo Fernando Calvo
Independent consultant on energy efficiency, Nicaragua
seifercalvo@gmail.com

Claudio Carpio
Fundación Bariloche, Argentina
claucarpio@gmail.com

Rita Cavaleiro de Ferreira
Strong Water Utilities in Jordan, GITEC-IGIP GmbH, Germany
rita.cavaleiro@gic-group.com

Javier Hueichapán Henríquez
Empírica Consultores, Jefe de proyectos de Sustentabilidad, Chile
jhueichapan@empiricaconsultores.cl,
javier.hueichapan@gmail.com

Welner Collejo Jerez
National Office for the Control of the Rational Use of Energy (ONURE), Matanzas, Cuba
welnerc@oc.une.cu, wcollejo1980@gmail.com

Marina Yesica Recalde
Consejo Nacional de Investigaciones Científicas y Técnicas (CONICET), Argentina
Fundación Bariloche, Argentina
mrecalde@fundacionbariloche.org.ar

Odón de Buen Rodríguez
National Commission for the Efficient Use of Energy (CONUEE), Mexico
demofiloente@gmail.com

Luis Enrique Manzano Villafuerte
Universidad de las Fuerzas Armadas (ESPE), Ecuador
luis.manzanovillafuerte@gmail.com, lemanzano1@espe.edu.ec

Contents

List of Figures

List of Tables

Preface

For any professional working in energy efficiency and energy policy in the Latin American region, it is clear that despite great progress in many areas, the energy efficiency gap in some productive sectors of the demand side is still very big. Energy efficiency actions, which are being evaluated worldwide as some of the most cost-effective options for decarbonizing these sectors, are still very difficult to implement in the non-industrial world. Different local circumstances and barriers (economic problems, lack of knowledge & information, among others) prevent actors from making investments.

Nevertheless, some decades ago, a very interesting strategy to promote energy efficiency practices began to be implemented. A strategy born in Switzerland and widely implemented in Germany with promising results has spread to other countries: the energy efficiency networks (EENs) for energy management systems (EMSs) or the Learning Energy Efficiency Network (LEEN) in Management Systems. These instruments seem to be an effective alternative to remove information, knowledge, and capacity barriers, which are clearly relevant in the Latin American Countries (LAC). In our region, Mexico has been the most relevant experience followed by Brazil, Argentina, Chile, Nicaragua, and Ecuador. When we look at these experiences as promising actions, we ask ourselves: What are the specific characteristics of LEENs in our region? How can the experience of regional leading countries be useful to other countries that would like to implement the methodology?

That was the main reason why we decided to write this book. Our aim was to contribute to disseminating the methodology to other (developing) countries of the LAC region and other developing regions of the world. We wanted to show both the main results and the challenges confronted by each country and, especially, how they managed to solve them. We were clear about what we wanted to show, but we needed local voices to tell the story. Who could tell better these (different) experiences? We agreed that we need to look for a group of local experts in the energy efficiency area who were directly involved in the implementation of the LEENs in each case. Their voices are expressed in the different chapters of this book.

This book examines different experiences for the industrial, services and commercial sector in seven LAC: Mexico, Argentina, Brazil, Ecuador, El Salvador, Colombia and Nicaragua, highlighting the relevance and role of enabling conditions, the role of key stakeholders, the main contribution of LEENs to the energy efficiency promotion and, especially, the key challenges confronted.

The book is written for those who work in energy efficiency and desire to broaden their knowledge, as well as those involved in the policy side of energy efficiency. We have presented a balanced view of the theoretical background and specific case studies.

We hope that the reader will enjoy a trip through the LEENs experiences in the Latin American region. We are convinced of the relevant role of energy efficiency actions and their huge potential to contribute to the economic performance of the sectors, and the potential of LEENs for this purpose. We hope that the lessons learned in these LAC will be of great interest to other countries in the developing world, particularly for the elaboration of public policies that facilitate both their promotion and their implementation in the socioeconomic sectors involved.

Claudio Carpio

Marina Yesica Recalde

Acknowledgment

The editors gratefully acknowledge the huge effort and great work done by the authors, who worked hard to consolidate a lot of local information in an easy-to-follow format.

Special acknowledgments to Noé Villegas Alcántar, former head of the National Energy Management Systems Program (PRONASGEN) in Mexico's industrial sector and promoter of LEEN training in both Mexico and Central America, for the fruitful discussions and information.

We also thank the Deutsche Gesellschaft für Internationale Zusammenarbeit (GIZ) for their permission to use the available information about the LEENs in Colombia.

We gratefully acknowledge the support of the Consejo Nacional de Ciencia y Técnica (CONICET) and Fundación Bariloche for providing us with the place to develop our research, and the colleagues who worked with us for the very constructive regular discussions about energy efficiency policy. However, we want to emphasize that the opinions expressed in this book are those of the authors and the editors and do not represent the views of their institutions.

Acronyms and Abbreviations

ACHEE	Agencia Chilena de Eficiencia Energética (Chilean Energy Efficiency Agency)
ADE	Agencia Danesa de Energía (Danish Energy Agency)
AENOR	Asociación Española de Normalización (Spanish Association for Standardization)
AMENEER	Asociación Mexicana de Empresas de Eficiencia Energética y Energía Renovable (Mexican Association of Energy Efficiency and Renewable Energy Companies)
AMEXGEN	Asociación Mexicana de Empresas de Gestión Energética (Mexican Association of Energy Management Companies)
AMEXCID	Agencia Mexicana de Cooperación Internacional para el Desarrollo (Mexican Agency for International Development Cooperation)
ANCE	Asociación de Normalización y Certificación (Standardization and Certification Association)
APF	Administración Pública Federal (Federal Public Administration)
BID	Banco Interamericano de Desarrollo (Inter-American Development Bank)
CADIN	Cámara de Industrias de Nicaragua (Chamber of Industries of Nicaragua)
CAMEXA	Cámara Mexicano-Alemana de Comercio e Industria (Mexican-German Chamber of Commerce and Industry)
CANACAR	Cámara Nacional del Autotransporte de Carga (Commission on Environmental Cooperation for North America)
CCA	Comisión de Cooperación Ambiental para América del Norte
CCI	Comité de Coordinación Interinstitucional (Interinstitutional Coordination Committee)

CEAGUA	Comisión Estatal del Agua Morelos (State Water Commission Morelos)
LaCEL	Comisión Ejecutiva Hidroeléctrica del Río Lempa (Executive Hydroelectric Commission of the Lempa River)
CFE	Comisión Federal de Electricidad (Federal Electricity Commission)
CGCERyEE	Comité de Gestión por Competencias en Energías Renovables y Eficiencia Energética (Management Committee by Competences in Renewable Energy and Energy Efficiency)
CONAE	Comisión Nacional para el Ahorro de Energía (National Commission for Energy Saving)
CONUEE	Comisión Nacional para el Uso Eficiente de la Energía (National Commission for the Efficient Use of Energy)—Mexico
DOE	Departamento de Energía de los Estados Unidos (U.S. Department of Energy)
DOF	Official Gazette of the Federation
ECOVES	Economía Verde y Energía Sustentable (Green Economy and Sustainable Energy)
EE	Energy Efficiency
EENs	Energy Efficiency Networks
EPE	Empresa Productiva del Estado (State Productive Enterprise)
FIRA	Fideicomisos Instituidos en Relación con la Agricultura (Trusts Instituted in Relation to Agriculture)
GHG	Greenhouse gases
GIZ	Deutsche Gesellschaft für Internationale Zusammenarbeit
GPAGES	Gerencia de Protección Ambiental y Gestión Energética de Pemex (Environmental Protection and Energy Management of Pemex)
GUE	Grandes Usuarios de Energía (Large Energy Users)
IPCC	Intergovernmental Panel on Climate Change

LEEN	Learning Energy Efficiency Network
Minambiente	Ministry of Environment and Sustainable Development—Colombia
Mincomercio	Ministry of Commerce, Industry and Tourism—Colombia
NDC	Nationally Determined Contributions
NORMEX	Sociedad Mexicana de Normalización y Certificación (Mexican Society for Standardization and Certification)
PAESE	Programa de Ahorro de Energía en el Sector Eléctrico (Energy Saving Program in the Electricity Sector)
PAT	Plan Anual de Trabajo (Annual Work Plan)
PEMEX	Petróleos Mexicanos (Mexican Oil Company)
PRONASGEn	Programa Nacional para Sistemas de Gestión de la Energía (National Program for Energy Management Systems)
PTB	Instituto Nacional de Metrología de Alemania (German National Institute of Metrology)
PyMEs	Pequeñas y Medianas Empresas (Small and Medium Enterprises)
RdA	Red de Aprendizaje / Redes de Aprendizaje (RdA Learning Networks)
SE	Secretaría de Economía (Ministry of Economy)
Semarnat	Secretaría de Medio Ambiente y Recursos Naturales
Sener	Secretaría de Energía (Ministry of Energy)
SHCP	Secretaría de Hacienda y Crédito Público (Ministry of Finance and Public Credit)
SMEs	Small and medium-sized enterprises
SNR	Sistema Nacional de Refinación (National Refining System)
TecNM	Tecnológico Nacional de México (National Technology of Mexico)
Toe	Tons of Oil Equivalent
UPAC	Usuarios de Patrón de Alto Consumo de Energía (High Power Consumption Pattern Users)

Introduction

In a 2023 article, the *Financial Times* highlighted the role of energy efficiency as "the hero in the decarbonization paths." Arguing that energy efficiency has contributed to the energy evolution worldwide, thanks to the improvement in the performance of the equipment, and that: "*...today, this need for energy efficiency is more pressing than ever, as the world looks to decarbonize industries and infrastructure. The good news is that the technology needed to significantly reduce energy consumption is already available today....*"[1] Rosenow and Eyre (2022) go in a similar direction, suggesting that energy efficiency will become even more relevant in a world of Net Zero, both for meeting climate goals and achieving other societal goals. For this to happen, the benefits of energy efficiency must be rethought.

The relevance of energy efficiency is well-known worldwide and, therefore, it has become a key topic in governmental agendas over the last decades. According to the IEA (2023) in June 2023, 46 governments participating in the "IEA's 8th Annual Global Conference on Energy Efficiency" endorsed the "Versailles Statement: The crucial decade for energy efficiency," agreeing to strengthen energy efficiency actions in line with a doubling of global energy intensity progress each year this decade to 2030. Moreover, according to the

[1]https://www.ft.com/partnercontent/abb/energy-efficiency-the-unsung-hero-on-the-route-to-decarbonisation.html.

Learning Energy Efficiency Networks: Applications and Challenges in Latin American Countries
Edited by Claudio Carpio and Marina Yesica Recalde
Copyright © 2026 Jenny Stanford Publishing Pte. Ltd.
ISBN 978-981-4968-82-9 (Hardcover), 978-1-003-77468-6 (eBook)
www.jennystanford.com

DOI: 10.1201/9781003774686-1

IEA's Government Energy Spending Tracker, since 2020, almost USD 700 billion were spent on energy efficiency investment support.

Beyond emissions and energy avoidance, energy efficiency has a lot of benefits. The currently known multi-benefit approach of energy efficiency highlights that energy can contribute to other development objectives. For instance, energy efficiency influences industrial productivity. Therefore, it may have other benefits, e.g., the increase in energy efficiency may reduce production costs and increase production yields per unit of input (IEA, 2014).

Nevertheless, even though many industrial energy efficiency actions have been demonstrated to be cost-effective; however, companies do not implement them, for different reasons, which include the lack of enabling environments and barriers. In many industrial cases, the former includes the uncertainty of regulations, lack of energy efficiency awareness, and the uncertainty of energy supply and market prices (Duzgun and Komurgoz, 2014; Stadelmann, 2017; Stede, 2017; Zhu and Chertow, 2017; Safarzadeh et al., 2020). Within the latter, in industry, there are economic and financial barriers (short payback time, difficulties of access to capital), capacity and knowledge barriers, and lack of awareness (Stede, 2017; Trianni and Cagno, 2012; Thollander et al., 2013; Safarzadeh et al., 2020). Similarly, Schlomann (2016) argued that the most important economic and non-economic barriers confronted by the industrial sector are related to the information and knowledge gap regarding the saving potential for a company and existing financial support programs, a fear of the negative impact of energy efficiency measures on the quality of processes, lack of own capital to make investments, high transaction costs of energy efficiency investments, and risk aversion in borrowing capital.

Therefore, it has been difficult for most countries to initiate the efficiency pathway in their industrial sectors. Usually, firms face different problems that prevent them from implementing these cost-effective efficiency actions. These market failures, behavioral problems, and barriers, which are usually context and time-dependent, require the implementation of energy policies to address them.

There are many different alternatives that can be implemented to remove these problems, and they depend on the barriers that need to be addressed. Within the alternatives of strategies to remove information, knowledge, and capacity barriers, **energy efficiency networks (EENs) for energy management systems (EMSs)** seem to be an effective alternative.

The EENs were developed in the late 1980s or early 1990s in Switzerland, supported by the Swiss Energy Agency, with the purpose of bringing together a group of companies with the aim of working collaboratively on the implementation of energy efficiency measures (Köwener et al., 2011; Rohde et al., 2015). Nearly a decade after the first EEN in Switzerland, this idea was transferred and adopted in Germany, wherein first-year results were promising enough to encourage the further creation of networks. Then, in 2006, the first concept of the **Learning Energy Efficiency Network (LEEN) in Management Systems** was introduced and performed during 2006–2009, based on five networks (Köwener et al., 2014).

By 2023, Germany, which is worldwide the most successful case of study, implemented nearly 4,000 projects for energy efficiency networks, with 3,243 companies registered, some of them running in parallel[2]. These networks have addressed very important results, such as around Euro 180,000 per year per company reduction in the costs associated with energy use; reduction in CO_2 emissions by 2.4% per year; and improvement in energy performance by 2.1% per year.

After the first years of the German experience, the concept of LEEN started to spread worldwide, especially with the idea of promoting management systems. The networks were introduced in Austria, China, Japan, and other developing countries. In concordance with its global leadership on energy efficiency networks, on behalf of Germany's G7 Presidency in 2015, the Federal Government and the business community worked to promote the concept of energy-efficiency networks, and this dialog has continued in the framework of the International Partnership for Energy Efficiency Cooperation (IPEEC). Furthermore, the German Cooperation Agency has acted as one of the most relevant funding agencies for the implementation of LEEN, at least in the case of Latin American Countries. In the Latin American region, Mexico has been the most relevant experience led by Brazil, Argentina, Chile, Nicaragua, and Ecuador.

A LEEN is a collaborative methodology between different actors pursuing the objective of optimizing the energy performance of their

---

[2]https://www.effizienznetzwerke.org/.

organizations through the exchange of experiences and the technical support offered by experts. In some countries, this objective was also extended to the implementation of energy management systems as a more comprehensive approach, beyond the identification measures. As argued by Stede (2019) networks operated through information, encouraging energy efficiency investment by providing companies with the knowledge and skills to effectively reduce energy consumption. The empirical evidence shows that the process can vary significantly from case to case in many aspects (financial scheme, number of meetings, number of participants, among others). However, there are some aspects in which these networks are similar. However, to succeed, LEENs are required to fulfill certain characteristics (Jochem and Gruber, 2007).

In this context, the ***Learning Energy Efficiency Networks: Applications and Challenges in Latin American Countries*** book, aims to present the fundamentals of Learning Energy Efficiency Networks and their contribution to the energy efficiency performance in the region. The book is based on existing experience in Latin America LEENs for EMSs in the industrial and service sectors, highlighting the relevance and role of enabling conditions, the role of key stakeholders and, especially, the key challenges confronted. It focuses on the recent experience of a group of LA countries: Mexico, Argentina, Brazil, Colombia, Chile, Cuba, Ecuador, and Nicaragua.

One of the main contributions and potentials of the book is that lessons learned in these Latin American Countries (LACs) can be of great interest to other countries in the developing world, particularly for the elaboration of public policies that facilitate both their promotion and implementation in socio-economic sectors.

This book is organized in two sections divided into seven chapters. The first section of the book is composed of theoretical chapters, which provide the academic contextualization of the cases of study. The second section includes 4 chapters, which address different cases of study of LACs that have implemented LEENs in the last decades. The methodology for the elaboration of these chapters has been discussed with different cases. In some cases, the chapters have been elaborated by local authors, and in other cases, they were based on available information and interviews. Finally, the last section includes the final reflections (Chapter 7).

Chapter 1 introduces the topic of energy efficiency its benefits, potential and the contribution of energy efficiency actions to the climate change mitigation policy. Then, the energy-efficiency gap is introduced and the main barriers and boundary conditions that explain the lack of energy investment are discussed. In a similar direction, it includes some academic perspectives on determinants of investment in energy efficiency and their links to the context conditions. The chapter ends with a presentation of the main policies and instruments used to remove those energy efficiency barriers and the energy gap, where the LEENs can play a relevant role.

Chapter 2 introduces Energy Efficiency Networks. An idea that was born in the late 1980s/early 1990s, in Switzerland, supported by the Swiss Energy Agency, and the Learning Energy Efficiency Networks highly supported worldwide. This chapter includes a brief description of the definition of the LEENs, the methodology for their implementation, the process, the participants, and their benefits.

Chapter 3 presents the case of the first and region's leading country developing Energy Efficiency Networks: the Mexican Case. This chapter has been developed by two local experts in energy efficiency, highly related to the implementation of LEENs, Odón de Buen Rodríguez, and Daniela Méndez Bellamy. The Mexican case is well-known in the region, not only because of the implementation of its networks but also for its appropriation of the methodology, updating some of the central points to the intrinsic characteristics of the country. This chapter is divided into two sub-chapters, the first one presenting the experience of industrial networks at the national level, and the second one discussing the experience of LEENs at the subnational level (municipalities).

Chapter 4 presents Industrial LEENs in Argentina, which initiated with one network in 2018 as a result of a triangular cooperation between the Mexican and Argentinean Governments, and German Cooperation for Sustainable Cooperation for Sustainable Development (GIZ), followed by five LEENs financed by the European Community, two additional ones financed by local governments and, more recently, four in small and medium-sized enterprises (SMEs) in the provinces of the Patagonian region of the country, one of which has been completed and the rest are just getting started. In this case, LEENs have included big and small industrial companies, in a country in which industrial consumption is responsible for 23%

of total final energy consumption and is highly dependent on the use of natural gas. LEENs seem to be, in this case, a good tool to remove the information and knowledge barriers faced by the industrial sector.

The case presented in Chapter 5, written by Rita Cavaleiro de Ferreira, is significantly different from the previous ones, as it addresses the experience in LEENs for the Urban Water Supply Systems in Brazil in the 2017–2021 period. This chapter presents the experience of the seven LEENs carried out within the German–Brazilian cooperation.

Chapter 6 briefly presents the experiences of LEENs in five additional LAC countries: Chile, Colombia, Cuba, Ecuador, and Nicaragua. The methodology of study used for the evaluation of these cases is different from the one in the previous chapters. In some cases (Chile, Ecuador, and Cuba), they were developed by local experts (Javier Hueichapán Henríquez in Chile; Luis Enrique Manzano Villafuerte for Ecuador; and Angel Alfredo Núñez Barreto and Welner Collejo Jerez for the Cuban case), while in others (Colombia), it was based on official information provided by the LEENs financing institution or (Nicaragua) developed based on an interview to Jairo Fernando Calvo.

Finally, Chapter 7 presents the final reflections and lessons learned from these cases of study.

Chapter 1

The Industrial Energy Efficiency Gap and the Barriers

Marina Yesica Recalde
Consejo Nacional de Investigaciones Científicas y Técnicas (CONICET), Argentina
Fundación Bariloche, Argentina
mrecalde@fundacionbariloche.org.ar

1.1 Industrial Energy Efficiency: Benefits and Potentials

Energy efficiency has become one of the main topics of interest worldwide, both for energy policy and climate change strategies. Its relevance has increased in the last decades as a mitigation strategy because empirical evidence has shown that supply-side initiatives (especially the increase in renewable energy share) will not be enough, per se, to reduce the environmental impact of the energy sector.

Learning Energy Efficiency Networks: Applications and Challenges in Latin American Countries
Edited by Claudio Carpio and Marina Yesica Recalde
Copyright © 2026 Jenny Stanford Publishing Pte. Ltd.
ISBN 978-981-4968-82-9 (Hardcover), 978-1-003-77468-6 (eBook)
www.jennystanford.com

DOI: 10.1201/9781003774686-2

The Sixth Mitigation Report of Working Group III of the Intergovernmental Panel on Climate Change (IPCC) has stressed that the relevance of climate change mitigation on projected GDP increases when both damages from climate change and adaptation costs are assessed. According to the report, the results from economic damages from climate change show that the global cost of limiting warming to 2°C over the 21st century is lower than the global economic benefits of reducing warming. However, if only mitigation cost is accounted for, then there are negative effects of mitigation on GDP growth, which varies significantly across regions depending notably on economic structure, regional emissions reductions, policy design, and level of international cooperation. In addition, the results show a misbalanced result in mitigation benefits if either climate damage is toward the low end of the range, or future damages are discounted at high rates (IPCC, 2022).

The above debate is highly dependent on the estimated cost of mitigation options. It is straightforward that mitigation options costing less than the opportunity cost of tCO_2eq are cost-effective. Figure 1.1 shows the potential contribution to net emissions reduction in 2030 (relative to a specified emission baseline) and the cost of the alternatives in different sectors and subsectors as estimated by the IPCC (2022). As can be observed, for the three most relevant sectors in terms of energy consumption (Buildings, Transport, and Industry), the costs for energy efficiency options (both from technological change or rational use of energy), are lower than the reference technology (in blue) or correspond to low-cost alternatives between USD 0 and USD 20 per tCO_2eq (yellow).

However, as mentioned by the IEA (2014), energy efficiency has a lot of benefits, apart from the always-cited reduction of emissions and energy avoidance. The currently known multi-benefit approach of energy efficiency highlights that energy can contribute to other development objectives, rather than the environmental one. In a similar direction, the IPCC (2022) argued that in most models, evaluated macroeconomic implications of mitigation co-benefits and trade-offs are not quantified comprehensively across the above scenarios and depend strongly on mitigation strategies.

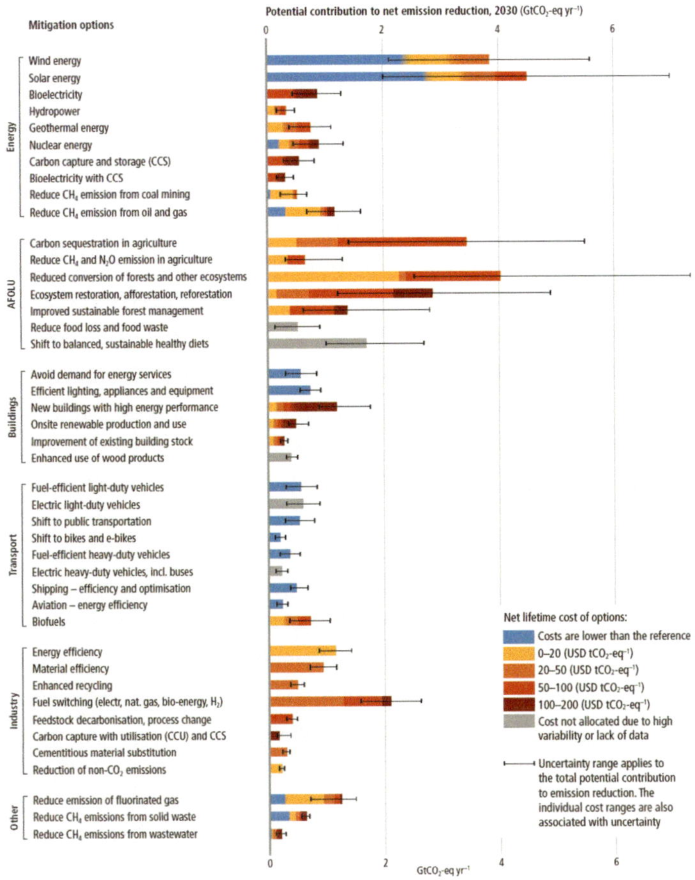

Figure 1.1 Overview of mitigation options and their estimated ranges of costs and potentials in 2030.

Source: IPCC (2022).

Indeed, regarding the role of energy in the multiple dimensions of development, the global promotion of energy efficiency has positive impacts in different directions, as shown in Figure 1.2. For instance, energy efficiency may have clear effects on energy poverty, and therefore, help in poverty alleviation. Energy poverty can be defined as “the state of being deprived of certain energy services or not being able to use them in a healthy, convenient, and efficient manner, resulting in a level of energy consumption that is insufficient to support social and economic development" (Bhatia and Angelou, 2015). Affordability of energy is crucial to analyze energy access, and therefore, energy poverty. It is important to note that affordability is not a unique variable; instead, affordability is a function of the package of energy, the price of energy, and the user's income level. As has been shown and discussed in several theoretical and empirical reports, the share of energy bills on total household expenditures is higher for the first deciles of income (poorest households) than in the last ones (richest households) (Jimenez and Yepez-Garcia, 2017). For example, Jimenez and Yepez-Garcia (2017) developed a cross-sectorial analysis based on the energy expenditures of 13 countries in the Latin American region. It was found that in all the countries in the poorest quantities, the share of energy expenditures is higher than the highest, with the situation being more inequitable for some specific fuels. For the poorest households of the region, electricity has the highest energy budget share (4.5% on average), followed by natural gas (3% on average).The implementation of energy efficiency actions (usually rational use of energy) is part of an energy poverty policy because the efficient consumption of energy reduces the unnecessary energy demand and reduces the energy bill, increasing the household disposable income.

In addition, energy efficiency actions in different sectors may reduce energy demand and consumption (in an efficient way), and therefore, increase energy security, which is particularly relevant for energy-importing countries. This aspect is especially important in the context of energy price volatility. For example, according to the World Bank, because of the Russia and Ukraine military conflict, oil prices increased up to 50% in 2022 compared to the previous year, while European gas prices were twice as high in 2022 as they were in 2021. This increase in energy prices and the shortage in the energy supply from Russia had a significant impact on the global macroeconomy. According to a KPMG's Chief Economist report published in 2023, simulation analysis performed found that a 10% rise in the price of oil together with a 50% rise in the price of gas because of Russia's invasion of Ukraine may imply a 0.5% reduction in the GDP of some of the larger European economies. While the impact of rising prices and spillover effects of energy prices because of the same conflict for economies outside Europe will be significant, the IMF's estimation suggests that a 10% rise in oil prices would decrease world GDP by 0.1–0.2%. Energy efficiency actions may constitute an important policy tool to attenuate the energy demand in final sectors, and therefore, reduce these negative impacts.

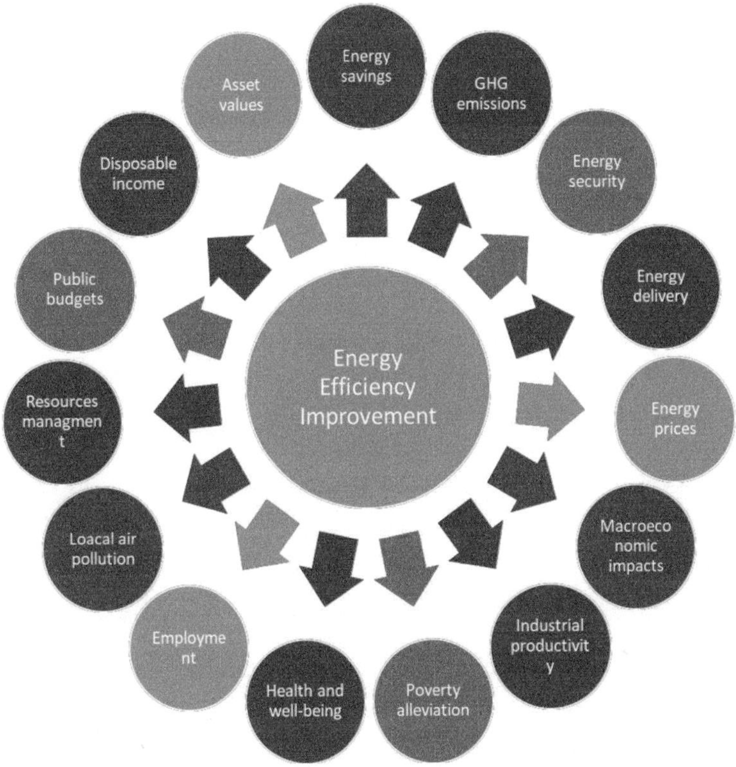

Figure 1.2 The multiple benefits of energy efficiency as stated by the IEA. *Source*: IEA (2014).

As shown in the previous figure, *industrial productivity* is one of the key positive impacts (benefits) that energy efficiency can bring to the economy. As stated in the Sixth Assessment Report of the Working Group III of IPCC, to get net-zero GHG emissions from the industrial sector coordinated action throughout value chains is needed to promote all mitigation options, including demand management, energy and material efficiency, circular material flows, abatement technologies, and transformational changes in production processes using low- and zero-GHG electricity, hydrogen, fuels, and carbon management (IPCC, 2022).

However, there are multiple ways in which energy efficiency influences industrial productivity, and therefore, it may have other benefits than energy use and emission reductions, e.g., the increase in energy efficiency may reduce production costs and increase production yields per unit of input (Figure 1.3). However, there are other effects of energy efficiency in the production process, such as improvements in operation and process reliability and reduction in equipment downtime, shutdowns, or system failures.

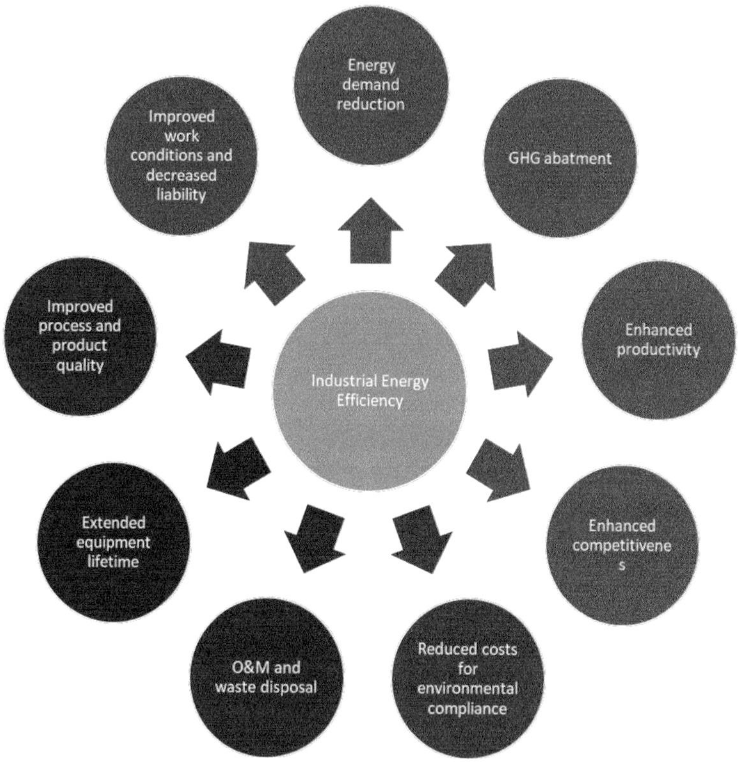

Figure 1.3 Direct and indirect benefits of energy efficiency in industrial sectors.

Source: own elaboration based on IEA (2014).

1.2 Contribution of Industrial Energy Efficiency to the SDGs

In 2015, 169 Nations of the world signed the document "The 2030 Agenda for Sustainable Development," in which they agreed on the need and relevance of achieving 17 Sustainable Development Goals (SDGs), which are an urgent call for action by all countries—developed and developing—in a global partnership. They recognize that ending poverty and other deprivations must go together with strategies that improve health and education, reduce inequality, and spur economic growth—all while tackling climate change and working to preserve our oceans and forests. The role of energy (and energy efficiency) in these SDGs is very important.

The role of energy consumption in socio-economic development has been widely discussed from theoretical and empirical standpoints. Energy constitutes a biophysical constraint to economic growth because every activity requires a minimum quantity of energy to be performed (Reister, 1987; Tahvonen and Salo, 2001; Recalde and Ramos-Martin, 2012; among others). Socioeconomic development requires a minimum of energy to be properly delivered (Recalde, 2017). The contribution of energy to sustainable development dimensions (economic, political, social, and environmental) is straightforward. The link between energy and socioeconomic development has also been highlighted by the international community by the recognition that it will not be possible to achieve the SDGs without an increase in global energy access. Access to appropriate, adequate, and affordable energy is closely linked to the achievement of several developmental objectives, including gender equality, education, health, food security, rural development, and poverty reduction. Therefore, the seventh SDG states, "Ensure access to affordable, reliable, sustainable and modern energy for all" (Figure 1.4). In this regard, the UN has defined four different targets, and the corresponding indicators will be useful to evaluate the performance of the regions and the world toward SDG7. In Table 1.1, targets and indicators directly related to energy efficiency have been highlighted.

Table 1.1 SDG7 targets and indicators

Target	Indicator
7.1 **By 2030, ensure universal access to affordable, reliable, and modern energy services**	7.1.1 Proportion of the population with access to electricity
	7.1.2 Proportion of the population with primary reliance on clean fuels and technology
7.2 **By 2030, increase substantially the share of renewable energy in the global energy mix**	7.2.1 Renewable energy shares in the total final energy consumption
7.3 **By 2030, double the global rate of improvement in energy efficiency**	7.3.1 Energy intensity measured in terms of primary energy and GDP
7.A **By 2030, enhance international cooperation to facilitate access to clean energy research and technology, including renewable energy, energy efficiency, and advanced and cleaner fossil-fuel technology, and promote investment in energy infrastructure and clean energy technology**	7.A.1 Mobilized amount of United States dollars per year starting in 2020 accountable toward the $100 billion commitment
7.B **By 2030, expand infrastructure and upgrade technology for supplying modern and sustainable energy services for all in developing countries, in particular least developed countries, small island developing states, and land-locked developing countries, in accordance with their respective programs of support**	7.B.1 Investments in energy efficiency as a percentage of GDP and the amount of foreign direct investment in financial transfer for infrastructure and technology to sustainable development services

Source: Inter-Agency and Expert Group on SDG Indicators (IAEG-SDGs).

Figure 1.4 The SDG7 and SDG9.

Source: https://www.un.org/.

According to the UN information data available in 2022, the global primary energy intensity—defined as the ratio of global total energy supply per unit of GDP—improved from 5.6 megajoules per dollar (2017 PPP) in 2010 to 4.7 megajoules in 2019. Since 2015, global energy intensity has improved by 1.6% per year on average, which is still short of the 3.2% annual rate now needed to reach SDG7.3.

Regarding the industrial sector, SDG9 states "*Build resilient infrastructure, promote inclusive and sustainable industrialization and foster innovation*" (Figure 1.4). The UN has defined eight different targets, and the corresponding indicators will be useful to evaluate the performance of the regions and the world toward SDG9. In Table 1.2, targets and indicators directly related to energy efficiency have been highlighted.

According to the 2022 information, global carbon dioxide (CO_2) emissions declined by 5.8% in 2020, or almost 2 billion tons—the largest decline since 1990 and almost five times greater than the 2009 decline that followed the global financial crisis. Despite the decline in 2020, global energy-related CO_2 emissions remained at 31.5 billion tons, which contributed to CO_2 reaching its highest average annual concentration in the atmosphere. In 2021, global energy-related CO_2 emissions rose by 6.0% to 36.3 billion metric tons, their highest-ever level, as demand for coal, oil, and gas rebounds with the economy.

Table 1.2 SDG9 targets and indicators

Target	Indicator
9.1 **Develop quality, reliable, sustainable, and resilient infrastructure, including regional and trans-border infrastructure, to support economic development and human well-being, with a focus on affordable and equitable access for all**	9.1.1 Proportion of the rural population who live within 2 km of an all-season road
	9.1.2 Passenger and freight volumes, by mode of transport
9.2 **Promote inclusive and sustainable industrialization and, by 2030, significantly raise the industry's share of employment and gross domestic product, in line with national circumstances, and double its share in least developed countries**	9.2.1 Manufacturing value added as a proportion of GDP and per capita
	9.2.2 Manufacturing employment as a proportion of total employment
9.3 **Increase the access of small-scale industrial and other enterprises in developing countries, to financial services, including affordable credit, and their integration into value chains and markets**	9.3.1 Proportion of small-scale industries in total industry value-added
	9.3.2 Proportion of small-scale industries with a loan or line of credit
9.4 **By 2030, upgrade infrastructure and retrofit industries to make them sustainable, with increased resource-use efficiency and greater adoption of clean and environmentally sound technologies and industrial processes, with all countries acting in accordance with their respective capabilities**	9.4.1 CO_2 emission per unit of value-added

Target	Indicator
9.5 **Enhance scientific research, upgrade the technological capabilities of industrial sectors in all countries, in particular developing countries, including, by 2030, encouraging innovation and substantially increasing the number of research and development workers per 1 million people and public and private research and development spending**	9.5.1 Research and development expenditure as a proportion of GDP 9.5.2 Researchers (in full-time equivalent) per million inhabitants
9.a **Facilitate sustainable and resilient infrastructure development in developing countries through enhanced financial, technological, and technical support to African countries, least developed countries, landlocked developing countries, and small island developing states**	9.a.1 Total official international support (official development assistance plus other official flows) to infrastructure
9.b **Support domestic technology development, research, and innovation in developing countries, including by ensuring a conducive policy environment for, inter alia, industrial diversification and value addition to commodities**	9.b.1 Proportion of medium and high-tech industry value added in total value added
9.c **Significantly increase access to information and communications technology and strive to provide universal and affordable access to the internet in least developed countries by 2020**	9.c.1 Proportion of population covered by a mobile network, by technology

Source: Inter-Agency and Expert Group on SDG Indicators (IAEG-SDGs).

1.3 Understanding the Industrial Investment Behavior in Energy Efficiency

Despite this well-known relevance and positive aspects, energy efficiency investments (in all economic sectors, but particularly in the industrial one) remain below the desired levels.

In the energy literature, this phenomenon has been mentioned as the energy paradox, or the **energy efficiency gap** (Jaffe and Stavins, 1994; Allcott and Greenstone, 2012; Gillingham and Palmer, 2014; Gerarden et al., 2017; Giraudet and Missemer, 2023, among others). As stated by Giraudet and Missemer (2023), this concept was initially developed in the late 1970s and 1980s and reached maturity in the early 1990s, referring to a suboptimal level of energy efficiency investment. The problem can equally affect the extensive and intensive margins of investment, i.e., produce too few and/or too small investments.

As argued by the IEA (2014), to stimulate investments in energy efficiency in industries, it is fundamental to understand what drives business investment decisions. As discussed below, governments spent a lot of money and time developing public policies with the objective of closing the energy efficiency gap. These policies are effective only when they are well-designed and focused, which, in turn, requires the comprehension of the determinants of the investments, and the barriers confronted by stakeholders when developing efficiency actions.

The first aspect to consider is that, depending on their size, companies have different stakeholders who interact and define the magnitude of the investment and their destiny, such as executives, managers, and technicians.

The combinations of the different determinants of investment in energy efficiency are shown in Figure 1.5, which was built following Recalde (2017). This diagram synthesizes the relation between institutional conditions (political and economic) and energy efficiency deployment, and its impact on investment (public and private). Three relations require special attention.

Firstly, the positive impact of institutional quality on economic performance and, consequently, on the performance of investments.

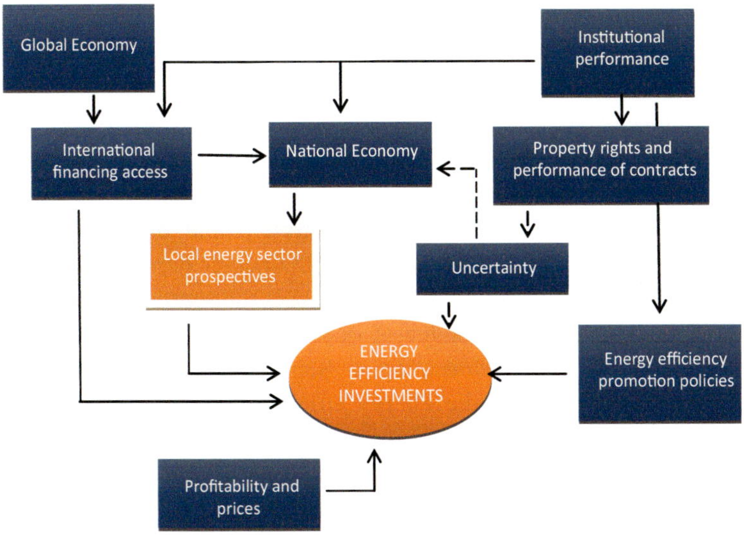

Figure 1.5 Determinants of investment in energy efficiency actions.

Source: Own elaboration.

Secondly, there is an indirect relationship between institutional quality and uncertainty, and between the latter and investments. The higher the institutional quality, the greater the possibility of agents following rules and conventions, and the lesser the uncertainty arising from their interaction. Less uncertain environments are more conducive to investments.

Thirdly, the higher the quality of political institutions, the greater their capacity to develop promotion policies. This aspect is very important, especially in less developed and developing regions dependent on governmental policies and programs. Additionally, energy prices play a significant role, as the lower the energy prices the lower the incentive to invest in energy efficiency options.

1.4 Border Conditions, Enabling Frameworks, and Barriers to Industrial Energy Efficiency

Many industrial energy efficiency actions have been demonstrated to be cost-effective; however, companies do not implement them, for different reasons. The literature is full of papers studying and evaluating the problems confronted by agents. This is part of the vast literature on barriers, which includes studies for specific sectors and regions of the world (Langlois-Bertrand et al., 2015; Sorrell, 2004; Sorrell et al., 2011; Triani et al., 2016; Yeatts et al., 2017; Vogel et al., 2015, among others). The studies have shown that barriers have different levels, ranging from problems affecting the whole socioeconomic system (border conditions), and other affecting various sectors and the energy sector to specific barriers affecting certain actions or instruments. 1.6 presents these categories.

The first and general levels of barriers are those elements belonging to the socioeconomic (national and international) system that frame all the energy efficiency actions—the border conditions. These elements set out the context of the policy and the energy efficiency actions and do not depend on the decision-making mechanism of the sector itself. The border conditions may be external (international oil market; negotiations in the framework of Climate Change, lack of access to certain technology; global economic conditions, international agreements) or internal (institutional structure, general macroeconomic aspects such as the foreign investment law, conditions with respect to the external sector, foreign trade, and local financial market). It is straightforward that these conditions are country and time-specific and affect efficiency actions and even policy results (it will be mentioned later).

The second level of problems is named enabling frameworks or enabling environments. They represent the relevance of an environment conducive to the implementation of energy efficiency actions, or for the implementation of certain instruments and specific policies. They are necessary but not sufficient conditions for the implementation of energy efficiency actions. Although in some cases, these aspects may be out of the scope of energy efficiency policymakers they should be somehow modified by national authorities before the implementation of other actions. Nygaard et al. (2015), based on definition from the Intergovernmental Panel on Climate Change (IPCC), define an enabling environment as: "*the entire range of institutional, regulatory, and political framework conditions that are conducive to promoting and facilitating the transfer and diffusion of technologies, including the country-specific circumstances that encompass existing market and technological conditions, institutions, resources, and practices, which can be subject to changes in response to government actions*". If these conditions are not present, the implementation of actions may be more complex. There are main areas that governments can influence directly or indirectly to modify the enabling conditions. There is a vast literature discussing enabling conditions and energy and climate change policies (Duguma et al., 2014; Recalde, 2017; Recalde et al., 2018; Zabaloy et al., 2019), as well as enabling conditions and technology transfer (Nygaard et al., 2015; Ockwell et al., 2015; Trærup et al.; 2015; Watson et al., 2015).

In many industrial cases, the lack of enabling frameworks including the uncertainty of regulations, lack of energy efficiency awareness, and the uncertainty of energy supply and market prices (Duzgun and Komurgoz, 2014; Stadelmann, 2017; Stede, 2017; Zhu and Chertow, 2017; Safarzadeh et al., 2020).

The distinction between these two categories is not so clear because some enabling frameworks can also be a boundary condition for the policy. Therefore, in many opportunities in the literature, these concepts have been used in both directions. Ellis (2010), for instance, stated that in some developing countries, there are many enabling conditions that influence energy efficiency investments, such as energy prices, fluctuations, political and economic instability, etc. However, these conditions have been classified either as border or enabling conditions according to different papers.

The third and probably the most relevant groups of problems are those known as specific barriers. They are defined as mechanisms that hinder investment in energy-efficient and economically profitable technologies, such as lack of knowledge of energy efficiency potential, high costs of financing, lack of funding alternatives, and lack of technical capacities to implement energy efficient actions, among others (Ruchansky et al., 2011). The literature on barriers to energy efficiency actions is wide (Brown, 2014; Langlois-Bertrand et al., 2015; Sorrell, 2004; Sorrell et al., 2011; Triani et al.; 2016; Yeatts et al. 2017; Vogel et al., 2015, among others) and several empirical studies evaluate specific barriers in particular sectors of the economy.

Specific barriers for each sector are country and time-dependent, and their intensity differs from one country to another because they depend on the market development, and policies implemented historically, among other aspects (Ruchansky et al., 2011). At the same time, specific barriers are dependent on the border conditions. In the case of the industrial sector, barriers can differ significantly among different countries, sizes of the company, and branches of economic activity. This is especially clear for the difference between barriers confronted by large industrial companies and small and medium enterprises (SMEs) within the industrial sector.

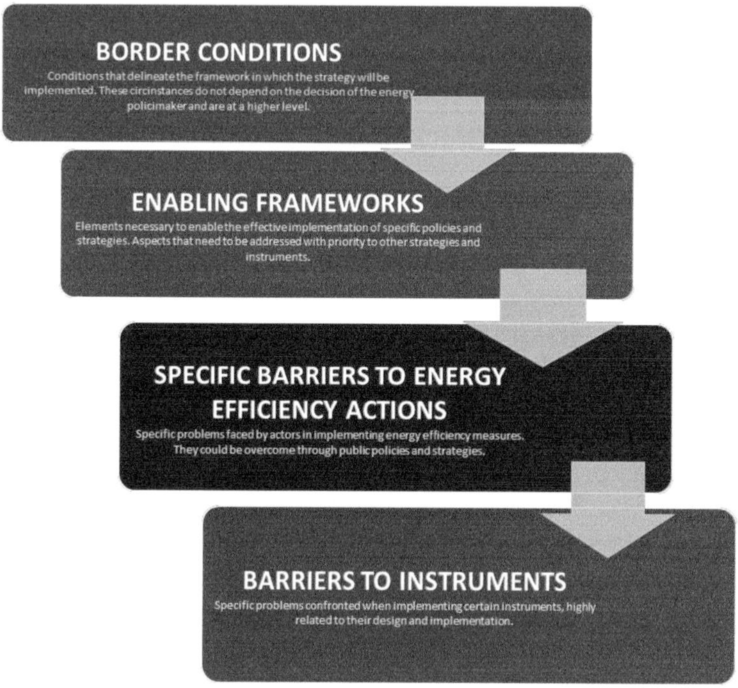

Figure 1.6 Definition and scope of different categories of barriers.

Source: adapted from Bouille et al. (2019a).

Within the industry, there are economic and financial barriers (short payback time, difficulties of access to capital); capacity and knowledge barriers, and lack of awareness (Stede, 2017; Trianni and Cagno, 2012; Thollander et al., 2013; Safarzadeh et al., 2020). Similarly, Schlomann (2016) argued that the most important economic and non-economic barriers confronted by the industrial sector are related to the information and knowledge gap regarding the saving potential for a company and existing financial support programs, a fear of the negative impact of energy efficiency measures on the quality of processes, lack of own capital to make investments, high transaction costs of energy efficiency investments, and risk aversion in borrowing capital.

However, as stressed by Wohlfarth et al. (2017) and Safarzadeh et al. (2020), usually information deficits (imperfect information) are the most relevant challenges in industrial sectors, and this may require the implementation of instruments to promote internal knowledge capacity-building, information gathering, and technical capacity-building. Anderson and Newell (2004) found that the information provided by industrial energy audits is highly relevant to reducing energy consumption.

1.5 The Role of Public Policy and Instruments

Closing this energy efficiency gap and removing the barriers requires the implementation of policies, actions, and instruments. As stated by Tanaka (2011), the evidence of great potential for cost-effective efficiency-derived reductions in industrial energy consumption and GHG emissions has encouraged governments to implement different policies and measures to improve energy efficiency in this sector. The Office of Energy Efficiency & Renewable Energy from the US Department of Energy argued that "*Energy efficiency policies and programs can help drive the implementation of projects that minimize or reduce energy use during the operation of a system or machine and/or production of good or service.*"

Governments can, therefore, mitigate barriers by designing and implementing well-targeted energy efficiency policies, programs, and instruments (Jacobsson et al., 2017; Kern et al., 2017; Rosenow et al., 2016; Grubb et al., 2017; Purkus et al., 2017; Rosenow et al., 2017; Liu et al., 2019).

It is straightforward that removing barriers requires, usually, more than one instrument. In the case of industries, empirical evidence shows that firstly awareness problems should be addressed through policies to identify energy efficiency potential and opportunities and increase the willingness for those investments. Then, removing information gaps requires providing information on measures, technologies, products, and financial benefits of energy efficiency to help enterprises make energy efficiency improvement decisions. Finally, technical capacity problems are usually the lack of internal capacities to identify and implement energy efficiency opportunities, as well as the lack of access to technical support (Carpio and Recalde, 2021).

There are multiple classifications of instruments, according to their nature and the category of the barrier they try to remove. Figure 1.7 shows a specific classification proposed by Tanaka (2011) for policy instruments in the industrial sector.

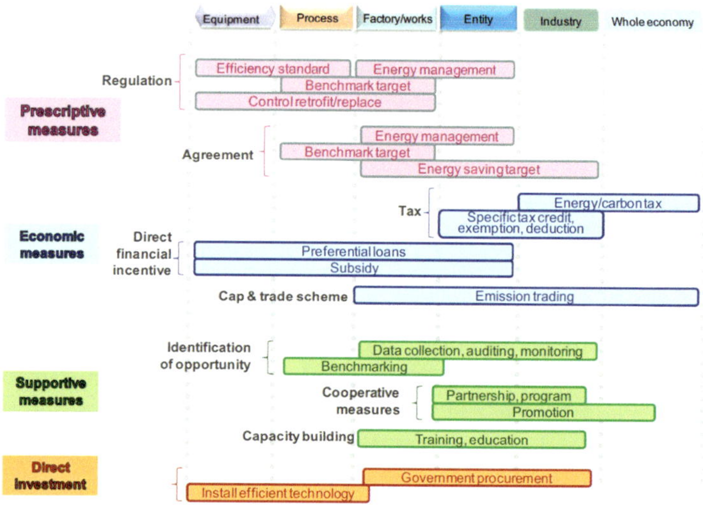

Figure 1.7 Coverage of policies in the industrial sector.

Source: Tanaka (2011).

Yet, the empirical evidence has shown that in some cases, well-designed instruments have not been successful in closing the gap (Zabaloy et al., 2019). Understanding the key factors affecting the energy policy is crucial to increasing the performance of the strategies and instruments.

This claims the recognition that country-specific aspects impact the process for the elaboration of the energy efficiency policy. Indeed, as argued by CEPAL/OLADE/GTZ (2003), the performance and success of any energy policy are the result of the combination of different factors: the international and national context; the instruments chosen and designed to overcome the existing barriers for the penetration of different technologies or practices; and the attitude of key stakeholders of the sector in reaction to the context conditions and to the implementation of the instruments. Policy instruments emerge depending on the political dynamics of each country, and these dynamics have an impact on the development of the policy instrument itself (Rosenow et al., 2016). As stressed by Estache and Kaufmann (2011), this means that for some policies, the concept of best practice can be quite misleading.

The relevance of both context/border and enabling environment is crucial for the evolution of energy efficiency strategies. While the former frames the context of policies, the latter establishes a proper environment for instruments to succeed. When missing, these enabling conditions may result in a bad performance of tools.

For example, according to the studies performed by IEA (2007) and Fuller et al. (2009) among others, Estache and Kaufmann (2011) concluded that the majority of policies failed due to the underestimation of the relevance of basic agency issues and agents reactions, such as lack of a careful evaluation and design of instruments, as the cost-effectiveness of them can vary across designs, agents, sectors, countries, time, and the macro context (including global financial liquidity and the fiscal situation of the governments concerned with energy efficiency), as well as a misunderstanding of the role and impact of financing, prices, enforcement and information, communication, and education.

Some enabling conditions are governmental commitment to energy efficiency actions, synergy with other public policies directly related to the topic, a clear institutional organization for energy efficiency, governmental capacity for monitoring and evaluation of public policies, non-existence of energy subsidies or price distortions, a good information system, and clear and stable regulatory framework for energy efficiency actions are some of the key conditions that facilitate the implementation of energy efficiency actions.

From this standpoint, the Learning Energy Efficiency Networks (LEENs) can be classified as a capacity-building instrument to remove both technical capacity and information barriers. A LEEN operates different activities, with constant technical support for its participants, increasing the technical knowledge of the industrial sector. As argued by Wohlfarth et al. (2017), one of the main advantages of LEEN is that they go "*beyond the sole provision of consultation, by qualifying companies to decide and act on implementing energy efficiency measures by delivering the individual practical know-how they need.*" In some specific cases, this process may generate critical information needed to implement an efficient energy management system. Other ways in which LEENs can help remove specific energy efficiency barriers are by raising awareness of energy saving potential, increasing transparency on energy use and energy data throughout the implementation of energy management systems and monitoring tools, and filling the gap in public and private financing options (Schlomann 2016).

The next chapter will discuss the worldwide emergence of LEENs and its relevance to the energy efficiency markets.

Chapter 2

Learning Energy Efficiency Networks

Claudio Carpio[a] and Marina Yesica Recalde[a,b]
[a]*Fundación Bariloche, Argentina*
[b]*Consejo Nacional de Investigaciones Científicas y Técnicas (CONICET), Argentina*
claucarpio@gmail.com, mrecalde@fundacionbariloche.org.ar

2.1 History of the Energy Efficiency Learning Network

The Energy Efficiency Networks (EEN) idea was born in the late 1980s or early 1990s, in Switzerland, supported by the Swiss Energy Agency, with the purpose of bringing together a group of companies with the aim of working collaboratively on the implementation of energy efficiency measures. This idea grew significantly and by 2014 more than 70 networks existed in Switzerland (Köwener, 2014; Rohde et al., 2015).

In 2006, a tax on the use of fossil fuels was implemented in Switzerland but those companies that have reduced their energy-related CO_2 emissions within the framework of a negotiated and mandatory target, and which undergo an annual evaluation, can be exempted from this tax. As stated by Jochem et al. (2016), the EEN

Learning Energy Efficiency Networks: Applications and Challenges in Latin American Countries
Edited by Claudio Carpio and Marina Yesica Recalde
Copyright © 2026 Jenny Stanford Publishing Pte. Ltd.
ISBN 978-981-4968-82-9 (Hardcover), 978-1-003-77468-6 (eBook)
www.jennystanford.com

DOI: 10.1201/9781003774686-3

has proved to be a useful and effective tool to reduce CO_2 emissions by reducing their energy consumption, and therefore, has helped companies to avoid the payment of the tax. Köwener et al. (2011) stated that since 2008, the CHF 12 tax has provided substantial support for the EEN.

Nearly a decade after the first EEN in Switzerland, this idea was transferred and adopted in Germany. The first energy efficiency network was established and operated in Hohenlohe between 2002 and 2006. The first EEN was initiated in 2002. It was composed of 17 companies of different sizes from a broad range of industrial and service sectors of the Hohenlohe region, in the southwest of the country. The first-year results of this network were promising enough to encourage the further creation of LEENs. Energy efficiency improvements in the first year were 1.4% (on a total energy consumption basis), with a reduction of 2.5% of CO_2 emissions across all participants. This pattern of efficiency was maintained during 2002–2004. The electricity generation productivity increased by 3% per year while carbon intensity reduced by 2.5%; total energy savings were 28 TJ, and in 2004, total energy costs saved were Euro 450,000 (Jochem and Gruber, 2007).

After this first experience, Germany, which is worldwide known as the origin of the LEEN and the most successful case of study, has implemented several pilot projects for energy efficiency networks, with differences in their design, and some of them running in parallel.

In 2006, the first concept of a Learning Energy Efficiency Network (LEEN) in Management System was introduced and performed during 2006–2009, based on five networks (Köwener et al., 2014). The LEEN Management System funded by the German Environmental Ministry or NKI Program was developed by the 30 Pilot Network Project during the 2009–2014 period, which was then extended in the 2014–2018 period within the "LEEN 100 Plus." Between 2009 and 2013, the German government financed 30 other Pilot Learning Networks (Rohde et al., 2015). The main objective of this pilot was to demonstrate the benefits of the energy efficiency networks and establish a management system. According to the results published by Rohde et al. (2015) and Köwenerer et al. (2015), 370 companies that participated in them achieved: (i) reductions in the costs associated with energy use by around Euro 180,000 per year per company; (ii) reductions in CO_2 emissions by

2.4% per year; and (iii) improvements in energy performance[1] by 2.1% per year. German experience has shown that companies that are part of a Learning Network get better results in terms of their energy performance compared to those companies that implement measures on their own (OECD / IPEEC, 2017).

In 2009, a new model was initiated, as companies specializing in the implementation of the energy efficiency networking methodology began to emerge. By 2013, the support of the German government for the financing was reduced and the networks began to be self-financed (IPEEC, 2016). Then the Pilot Project Marie on energy efficiency networks for SMEs was introduced during the 2012–2016 period (Durand et al., 2018).

In 2014, the Pilot Project of 30 networks was extended to 100 Energy Efficiency Networks, all of them using the LEEN standard (Schlomann, 2016). The National Action Plan on Energy Efficiency (NAPE) published by the German federal government adopts the idea of the networks and sets the goal of implementing a total of 500 networks by 2020 (Rohde et al., 2015; Schlomann, 2016).

In December 2014, a new Energy Efficiency Network Initiative was launched. For this purpose, the German Federal Government signed an agreement with business associations and organizations on the nationwide introduction of energy-efficiency networks. The alliance aims to set up approximately 500 new networks by 2020. The aim of these networks is to make an important contribution toward boosting energy efficiency in industry, crafts, trade, and commerce[2]. The aim was to save up to 75 PJ of primary energy or 5 million tons of GHG emissions. As stated by the Federal Government, the Energy Efficiency Networks Initiative has been extremely successful in supporting companies in exchanging information and implementing measures to optimize energy efficiency, to the extent of being remarked as the most successful instrument of the NAPE, even though the 500 networks originally announced by industry have not been reached. According to governmental information, the efficiency measures taken by the companies within the 295 networks

---

[1]*Energy performance* of an industry is defined as the amount of *energy* actually consumed (or estimated to be necessary) to meet the different needs for their processes and production.

[2]https://www.bmwk.de/Redaktion/EN/Artikel/Energy/energy-efficiency-networks-initiative.html.

will contribute to the achievement of the CO_2 savings target by the end of 2020 (5 MM)[3].

The energy efficiency network initiative launched in December 2014 will be continued from January 2021 to December 2025, as the energy efficiency and climate protection network initiative, with the aim going beyond increasing energy efficiency in industry, trade, commerce, trade, and the energy sector, and including climate protection, energy transition, and sustainability. As stated on the web page of the initiative, the aim is to have 300–350 new networks initiated by the end of 2025, with a total saving of 7–9 TWh of final energy and 5–6 MM tons of GHG. This "new version" of the network, the Energy Efficiency and Climate Protection Networks Initiative is supported by the Federal Ministry of Economics and Climate Protection (BMWK) and the Federal Ministry for the Environment (BMUV) as well as by 21 organizations from German industry and supported by other cooperation partners. Registered networks and participants in the German Energy Efficiency Network Initiative by 2023 are shown in Figure 2.1.

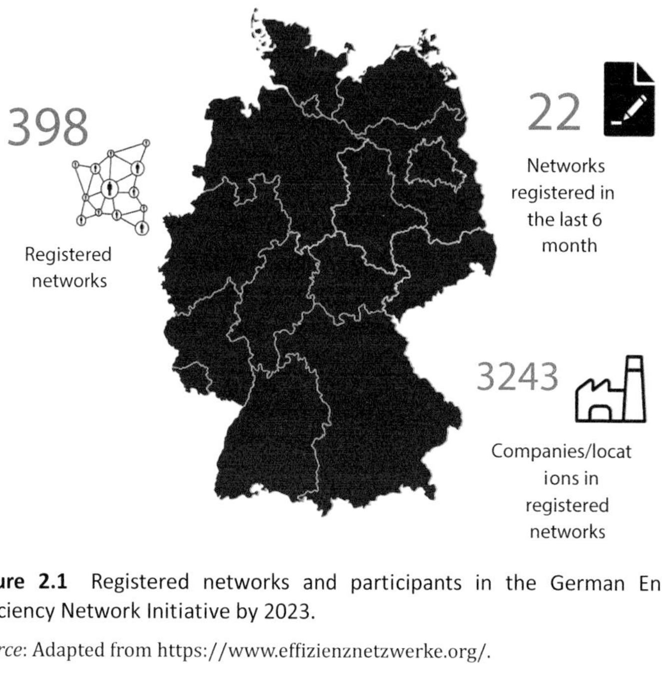

Figure 2.1 Registered networks and participants in the German Energy Efficiency Network Initiative by 2023.

Source: Adapted from https://www.effizienznetzwerke.org/.

[3]https://www.effizienznetzwerke.org/esgehtweiter/.

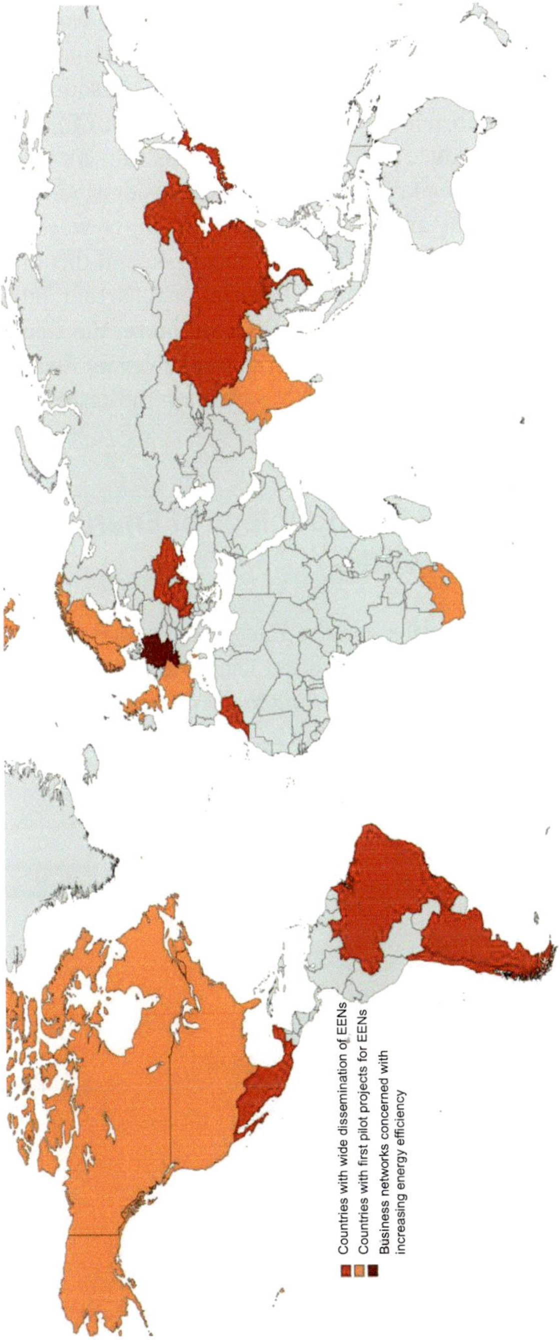

Figure 2.2 Situation of LEEN worldwide by 2022.

Source: own elaboration[4].

[4]https://www.mapchart.net/world.html.

In 2010, the concept of LEEN started to spread worldwide (Figure 2.2), especially with the idea of LEEN to Management Systems, and in 2012, the first networks were introduced in Austria (ILL-Werke, ENAMO), China (together with SGCC), Japan, and Brazil. In concordance with its global leadership on energy efficiency networks, on behalf of Germany's G7 Presidency in 2015, the Federal Government and the business community worked to promote the concept of energy-efficiency networks, and this dialog has continued in the framework of the International Partnership for Energy Efficiency Cooperation (IPEEC). Furthermore, the German Cooperation Agency has acted as one of the most relevant financing sources for the implementation of LEEN, at least in the case of Latin American Countries.

2.2 Definition and Methodology of Energy Efficiency Networks

There is no unique definition for the Energy Efficiency Networks or Learning Energy Efficiency Networks. These EENs or LEENs are usually redefined and adapted according to the different contexts of countries that implement them (Figure 2.3).

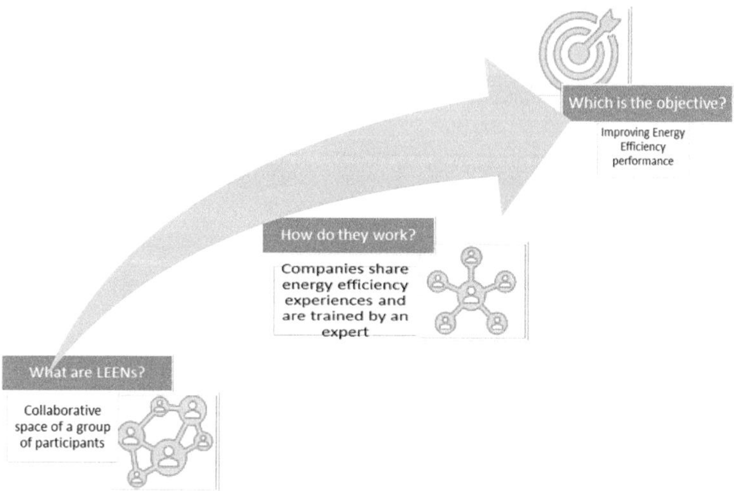

Figure 2.3 Definition of the LEENs.

Source: Own elaboration.

Essentially, the LEEN is a voluntary agreement of a small group of companies, usually between 10 and 15 firms, with the common objective of reducing energy consumption. These networks constitute a collaborative space in which participants interact during a period, exchanging experiences, problems, and opinions. The EEN is based on joint work and therefore in a co-construction of knowledge, based on the interaction of colleagues from different companies supported by experts and a moderator. As argued by Stede (2019), networks operated through information, encouraging energy efficiency investment by providing companies with the knowledge and skills to effectively reduce energy consumption.

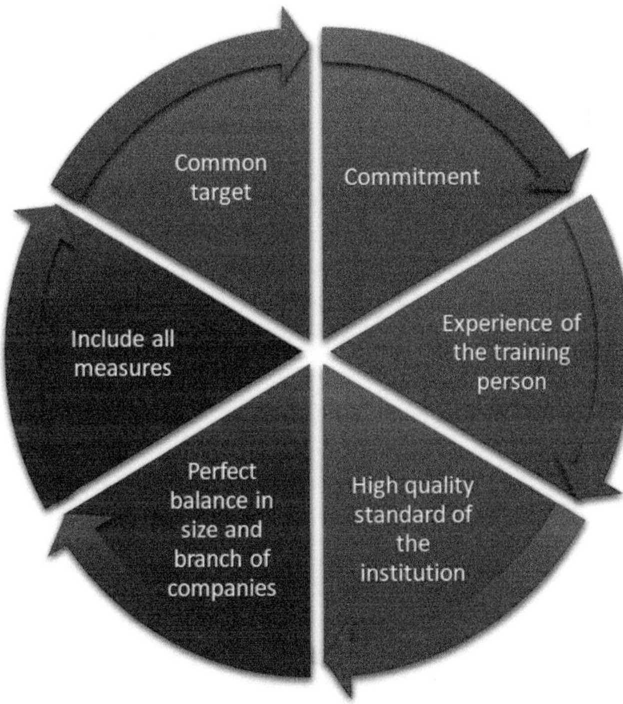

Figure 2.4 Key characteristics of the LEENs.

Source: Own elaboration.

To succeed, LEENs are required to fulfill certain characteristics (Jochem and Gruber, 2007) (Figure 2.4). Firstly, commitment from the representatives of the companies is crucial; this means the dedication of technical professionals as well as top management personnel of the company. Secondly, the role of the training person is highly important: he or she should be a professional senior engineer with a high knowledge of the production sites. Thirdly, the institution organizing the network should operate at a high professional standard. This is because the quality of the organization of the meetings, the information provided before, during, and after the workshops, and other logistical aspects are crucial to gathering the interest and involvement of participants. Fourthly, the companies participating should not be either too heterogeneous in size or branch or too homogeneous to avoid competition. Fifthly, all measures (even small measures) need to be considered in a systematic inventory of potential actions. Finally, a key convincing factor in some companies seems to be the common target-setting, which leads to the inclusion of some small measures, which otherwise would not have been considered.

A good performance of a network usually requires some national conditions. Jochem and Gruber (2007) stressed that some external boundary conditions may be important to establish a successful learning network. According to the authors, in the Swiss case, the role of the CO_2 Law was crucial as an incentive for companies: companies desired to get involved to avoid the CO_2 surcharge at a minimum cost.

In the first chapter, the industrial energy efficiency barriers were addressed. As stated before, the energy efficiency gap is mainly explained by the existence of problems/barriers confronted by private stakeholders. These barriers or market failures demand state intervention to be overcome. LEENs or EENs can be used, and have been indeed used, by governments as a policy instrument designed to remove the following barriers: economic or market barriers, funding barriers, institutional and regulatory barriers, information barriers, technological and capacity barriers, and cultural or awareness barriers.

Although these networks work on a voluntary basis, they are often incentivized by existing regulatory and policy frameworks even though they can also exist without government involvement (OECD / IPEEC, 2017). Usually, firms participating in the LEEN come from different economic sectors and are typically all from a specific region; although in some cases, there are sectorial networks

(Stede, 2019). As firms come from different sectors, LEEN focuses on efficiency improvements in cross-cutting technologies, such as lighting or process heat (Jochem et al., 2010).

2.3 The Process for the LEENs/EENs

As shown in Figure 2.5, the LEEN methodology allows participants to achieve a common goal through the exchange of experiences through meetings, courses, workshops, seminars, and webinars, among other activities, with constant technical support.

This methodology allows knowledge to be shared vertically and horizontally. In the first case, participants are trained and coached by experts and consultants, and they receive technical accompaniment. In the second case, horizontal learning is possible thanks to the interaction between participants.

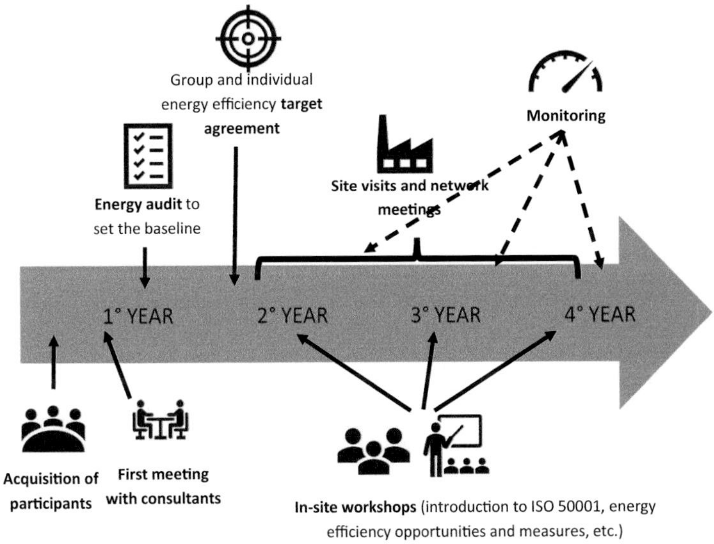

Figure 2.5 Schematic process for the LEENs.

Source: Own elaboration based on IPEEC 2017 Duran and Damian 2019.

The empirical evidence shows that the process can vary significantly from case to case in many aspects (financial scheme, number of meetings, among others). However, there are some aspects in which these networks are similar. The first is selecting the companies participating in the process (8–12 companies). It is crucial that the companies involved are committed enough to the objective of the network and allow the consultant or professional of the LEEN management team to perform the energy audits to evaluate the energy-saving potential of the companies. Participating companies in a LEEN must then set energy efficiency targets, both for individuals and groups. Then the LEEN process proceeds, with network meetings, site visits, and workshops in which capacity is built in different ways (technologies, organizational measures, ISO 50001, among others). Finally, the monitoring phase starts in which companies evaluate their performance.

ISO 50001

ISO 50001 is an international standard that consists of the implementation of an Energy Management System in the organization in which it is being applied, thus incorporating energy as a cross-cutting and fundamental element in strategic decision-making. While it shares the essence of what ISO 9001 (Quality Management System) and ISO 14001 (Environmental Management System) are, unlike them, this standard requires the organization to define and achieve improvements in its energy consumption, demonstrating its commitment.

The implementation of an EMS based on ISO 50001 identifies important opportunities to establish, implement, maintain, and continuously improve an energy management system. The key elements of ISO 50001 seek to identify the roles, responsibilities, and training needs that add value to production processes from an energy point of view. The implementation of ISO 50001 guarantees companies' compliance with their energy policy.

2.4 The Participants

One of the key lessons learned throughout more than 30 years of experience around the world on LEENs or EENs is that they must be adapted to the specific context of the countries. This also counts for the stakeholders involved in the process. However, there are at least four relevant stakeholders with different roles that are needed throughout the process. The EENs' success is built on their joint work, and therefore a clear definition of their roles and responsibilities is crucial. One of the most critical requirements is to hold a shared vision for objectives and methodologies. For instance, implementing partners and project carriers need to appreciate and understand the EEN as a service concept being delivered to their interest groups (OECD / IPEEC, 2017).

The learning process is, therefore, twofold. On the one hand, a LEEN seeks that the learning of the participants is vertical, by the experts, the consultants, and the technical accompaniment. On the other hand, it is equally relevant to achieve horizontal learning, that is, to be given thanks to the interaction between the other participants, each a specialist in their area of work. In LEEN, there are four fundamental actors.

Firstly, the network operator/initiator initiates the network and selects the members. Its main tasks/responsibilities are to motivate the emergence of the network, identify potential participants, and invite them to join it. Then, once the LEEN is running the network operator must follow the process. Different actors can play this role, and this depends on different country-specific aspects, such as the origin of the network, financial aspects, and other characteristics. Some of these stakeholders include companies, utilities, associations, chambers of commerce, service providers, and energy agencies, among others.

Secondly, the moderator organizes moderates and follows up meetings. The LEEN methodology is based on collective work performed during workshops, and then, the work of the moderator becomes critical. The main task of the moderator is to maintain synergy and exchange among participants, inducing them to share experiences, problems, and objectives. A very important goal of his/her work is to encourage the group to agree on a global energy efficiency performance target for the EEN and ensure its operation and monitoring. Moderators usually have a battery of participative techniques and methodologies that can be used to promote participation and collective work. The moderator role can be performed by different actors; however, usually, there are people specialized in such work.

Thirdly, as part of the work performed during the LEEN, companies involved in the process receive a technical expert or energy consultant who helps them to identify measures and give advice on energy-saving targets. This task is performed in different energy audits. The role of the energy consultant is, indeed, to accompany the participant of the LEEN throughout the whole process. He/she conducts technical workshops, conducts technical visits in the plant, carries out energy diagnostics, and determines the energy efficiency baseline of each participating company, among others. The people in charge of this activity must be highly qualified and could include an energy consultant, internal staff, external energy contractor, or a member of a utility.

Finally, the participants are, obviously, the most important actors in the LEEN. They are the reason why the network exists. In this case, the key point is to have participants committed to the achievement of both joint individual goals. This demands active participation providing information (technical and economic), attending workshops, participating in exercises complying with commitment and seriousness to the implementation of the activities established in the network. Participants regularly are representatives of companies, industries, institutions, or organizations, and this depends on the definition of the scope of the LEEN.

The network can be seen as a tree (Figure 2.6), for which the operator or initiator is the brain, or its roots, as it provides the initial idea or the origin and is responsible for its future evolution. As any tree requires the joint support of the sun and the rain, a LEEN demands the work of the energy consultant/expert and the moderator. Both are responsible for the networks' day-to-day evolution and development. Two additional stakeholders act in the direction of improving the LEEN: the accompanying entities (responsible for supporting and verifying information among participating companies); and the support entities (organizations and institutions that have already collaborated in the development of other Learning Network experiences). The role of these entities, which may or may not be present, is to fertilize the soil in which the networks grow, and to simplify the task for the rest of the members. The last component of this tree is the participants.

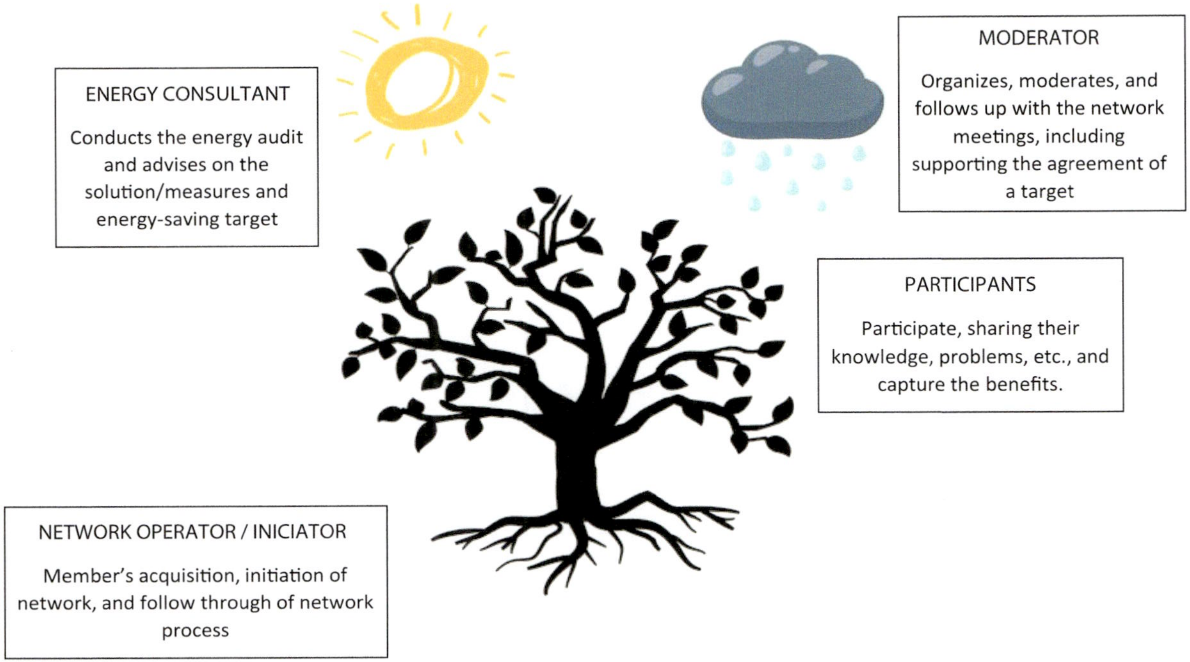

Figure 2.6 Graphical representation of the LEEN, participants, and their role.

Source: own elaboration.

2.5 Main Benefits of LEEN

The empirical evidence, especially in the countries with more experience in these networks (Switzerland, Germany, China, Mexico, among others), shows that besides energy efficiency performance, there are several co-benefits or advantages for companies participating in a Learning Network to implement an EMS. Most of these benefits are directly related to the industrial co-benefits of the energy efficiency mentioned above. Some of the advantages of EEN implementing EMS are:

- Transaction costs on topics of interest to participating companies are reduced. For example, energy saving, energy efficiency, cogeneration, use or applications of renewable energies, and suppliers, to name a few.
- The transfer of information and experiences on a specific topic is promoted and disseminated, and you learn to document success stories and best practices.
- Greater commitment is generated and the motivation of participants and their companies in energy issues increases.
- Improves environmental performance by increasing energy efficiency and reducing the environmental impacts of companies and industries.
- Learning networks are low-cost options since the services of specialized consultants and moderators of the networks are split among all participants.

Firstly, companies achieve a reduction in their production costs through lower energy costs, cogeneration, use or applications of renewable energy, and better management of suppliers and the supply system, among others. The changes in the operating routing usually have a reinforced effect on the improved performance of the company.

Secondly, the capacity of the company's workforce is strengthened through information and knowledge transfer of experiences and learning how to document success stories and best practices.

This rise in capacity and knowledge increases the innovative ideas of the companies and the reaction of their technology producers.

Thirdly, because greater engagement is generated, the motivation of participants and their companies on energy issues increases.

Fourthly, environmental performance is improved through lower energy use and better systems management.

Fifthly, through sharing the costs of the specialized consultants and the network moderator, the companies learn the benefits of cooperation among themselves.

Many of these benefits are possible thanks to the collaborative dynamic of the LEENs. As found by Chassein et al. (2018) and will be stressed in the following chapters, the cooperation within the network acts in the direction of increasing the company's commitment to addressing goals and targets, and the horizontal flow of information is key to the performance.

Chapter 3

Energy Management Experience in Mexico

Odón de Buen Rodríguez[a] and Daniela Méndez Bellamy[b]
[a]*National Commission for the Efficient Use of Energy (CONUEE), Mexico*
[b]*Senior international consultant*
demofiloente@gmail.com, dmendez.bellamy@gmail.com

3.1 The LEEN in the Mexican Industrial Sector

3.1.1 Background

Mexico initiated institutional energy-efficiency activities in 1989 with the creation of the National Commission for Energy Savings (CONAE), as a technical secretariat coordinating the efforts of several ministries and national energy companies to improve Mexico's energy efficiency. CONAE's specific mandate was to provide technical assistance on energy efficiency to all sectors of the economy, and its budget came from contributions by institutional members of the commission.

Learning Energy Efficiency Networks: Applications and Challenges in Latin American Countries
Edited by Claudio Carpio and Marina Yesica Recalde
Copyright © 2026 Jenny Stanford Publishing Pte. Ltd.
ISBN 978-981-4968-82-9 (Hardcover), 978-1-003-77468-6 (eBook)
www.jennystanford.com

DOI: 10.1201/9781003774686-4

To support these efforts, particularly in the industrial sector, CONAE had the assistance of international organizations, such as the European Commission, that supported a three-year program that included energy audits in a variety of industries, elaboration of technical guides by sectors, a national evaluation of cogeneration potentials, and the creation of regional offices in industrial chambers to provide technical assistance.

In 1995, the establishment of a working group between CONAE and PEMEX, the national petroleum company (which was the greatest contributor to CONAE's budget until 1998), gave place to the first formal attempt to organize a type of energy management system (EMS) in an industrial environment that included the creation of the group, the integration of internal committees within the different areas in PEMEX, data collection on energy use and product outcomes, the systematic identification of potential energy conservation actions, the definition of energy performance indexes, the setting of goals, and the evaluation of impacts, among other elements, that are the characteristics of present-day EMS (De Buen et al., 2003).

Also, in 1998, in response to budget restrictions in the public sector, CONAE designed and implemented a program to reduce energy use in federal public buildings, which included, also under a scheme similar to a large EMS, the creation of energy-efficiency committees in every ministry, data collection on energy use and size of buildings, standardized energy audits, the definition of energy performance indexes, yearly goals, work plans by each committee and a coordinating group led by CONAE, and annual recognition of the most relevant committees, among other elements of an EMS.

Later, from 2001 onward and based on the PEMEX experience and scheme, CONAE collaborated with national industrial chambers (steel, cement, and paper) in the form of specific working groups that replicated some of the practices in the PEMEX-CONAE collaboration.

In 2008, a new law (The Law for the Use of Renewable Energy and the Financing of the Energy Transition) transformed CONAE into CONUEE (National Commission for the Efficient Use of Energy). They gave it the mandate to collect yearly data on the energy use and energy efficiency practices of México's largest energy users in the industrial and transport sectors. In this way, a new program was established for these large energy users known as UPAC for its acronym in Spanish.[1]

3.1.2 Energy Management Systems (EMS) as Part of CONUEE's Programs

In 2013, a new federal administration renovated most of CONUEE's programs, including those related to the industrial sector, and incorporated within its general lines of action, for the first time and in an integral way, the promotion of EMS, particularly within the programs and activities related to the UPAC.

This was recognition of the development and integration of EMS in many countries as a well-established good practice with an international standard, ISO 50001, which made it possible to have a globally recognized method for the implementation of EMS.

An EMS is a methodology to achieve sustained and continuous improvement of energy performance in companies in a cost-effective way. The implementation of an EMS should not be understood as an objective but mainly as a powerful means to the objective of improving energy performance based on the results of the actions implemented throughout the system.

Understood in this way, the effectiveness of an EMS depends largely on the commitment and availability of all the actors involved in the company to manage the consumption and efficient use of energy, in addition to making the necessary changes in the day-to-day to facilitate these improvements. It also emphasizes that accomplishing efficient use of energy should not be done as an isolated effort, but under programmed planning in which all the levels involved in the company intervene, EMS being the best way to achieve it.

Thus, in 2013, CONUEE began with the promotion for the implementation of EMS in the UPAC and facilities of national energy companies and the First International Forum on EMS took place in México City.

In 2014, CONUEE went further and launched the **National Program for Energy Management Systems (PRONASGEn)**, which

[1]DOF, 2008. LEY PARA EL APROVECHAMIENTO DE ENERGÍAS RENOVABLES Y EL FINANCIAMIENTO DE LA TRANSICIÓN ENERGÉTICA. https://www.cre.gob.mx/documento/3870.pdf.

aimed to break barriers to harness their energy-efficiency potential by supporting medium and large energy users in the development of their capacities to raise their competitiveness through the sustainable use of energy by integrating EMS to their internal practices (CONUEE, 2019).

Some of the barriers identified were:

- Lack of allocation of resources (human, material, financial, and technological) within industries to implement a comprehensive energy efficiency program.
- Those responsible for production and maintenance at companies do not have time allocated for energy management, as their priorities are production, maintenance, and avoiding stoppages in processes.
- Lack of knowledge of the potential benefits and feasibility of good practices for energy efficiency.
- Lack of certified consulting companies that carry out energy efficient projects under international standards.

This initiative coincided with the international interest in the promotion of EMS and ISO 50001, and a number of international bodies offered their collaboration in CONUEE's initiatives, particularly German cooperation through the Deutsche Gesellschaft für Internationale Zusammenarbeit (GIZ) and the German Institute of Metrology (PTB, for its acronym in German); the Danish Energy Agency (ADE) from the Kingdom of Denmark; and the North American Commission on Environmental Cooperation (CEC), with the close collaboration of the United States Department of Energy (DOE).

The actions of PRONASGEn were opened to all energy users in the country, regardless of sector, size, or type of fuel used, but with specific and differentiated support according to the activities developed jointly with different allies. The global target was to implement and monitor at least 10 EMS each year.

The following lines of action were established for the execution of PRONASGen:

- Promote the development and supply of courses and materials to train human resources in the design, instrumentation, and operation of EMS.

- Promote and facilitate the link between energy users and consulting companies, educational institutions, research centers, and certification bodies for the development of EMS.
- Promote and support the training of human resources in the matter.
- Facilitate access to tools and information that allow the design and implementation of EMS in facilities.
- Facilitate the linking of energy users with international cooperation mechanisms and/or schemes that support the financing and development of actions in the matter.

To attract participants, the program offered technical assistance provided by CONUEE and international organizations, such as:

- Training of qualified staff on energy efficiency and EMS.
- Technical support for the implementation of the EMS, offered by national and international experts in accordance with ISO 50001 implemented.
- Energy efficiency potential identified.

An Inter-Institutional Coordination Committee (ITC) was established as a governance body for the formulation, implementation, and follow-up of PRONASGEn. The ITC was composed of the Director-General of CONUEE, who presided over it; representatives of the Ministry of Energy (SENER), Ministry of Environment and Natural Resources (SEMARNAT), Ministry of Finance and Public Credit (SHCP), and Ministry of Economy (SE); and representatives of the business sector, academic institutions, and international cooperation agencies supporting some of the program's activities.

The ITC met annually with a varied participation of representatives from the public, private, and social sectors.

3.1.3 Learning Energy Efficiency Networks (LEEN) in PRONASGEn

A central activity of PRONASGEn has been promoting, creating, and operating the LEEN for a diverse and broad group of actors from various sectors and areas, including the public (federal and municipal) and the private (large, medium, and small companies) across several sectors (industry, buildings, transportation, and municipal services).

Before the promotion of the LEEN model and strategy, a discussion took place in CONUEE around the concept of communities of practice to integrate and share knowledge by practitioners and to share success stories. Some steps in that direction included the creation of accounts on social media (Twitter) aimed at specific themes (buildings, transport, industry, standards, and municipal services) and energy end users. This led to conversations with GIZ, which proposed the consideration of the LEEN model for Mexico.

The first steps for the formation of LEEN in Mexico involved a series of meetings between GIZ and CONUEE, aimed at achieving appropriation of the concept and seeking its implementation in the country. During the second semester of 2014, a technical visit to Mexico from German experts had the objective of explaining in detail the concept of the LEEN and its methods of operation in Germany, and the possibility of implementing LEEN in Mexico was proposed (GIZ, 2016).

At the beginning of 2015, CONUEE and GIZ defined a work plan to start the first pilot project in Mexico on EMS-specific LEEN, which implied, on the one hand, an open invitation to the so-called UPACs to join and participate in the LEEN. On the other hand, to initiate a process for the development of capacities of both moderators, technical experts, and the GIZ and CONUEE staff, in the structuring and functioning of a LEEN, and on the EMS.

CONUEE then formalized the invitation to UPACs to participate in a workshop to explain in detail the characteristics of the LEEN, the roles and responsibilities of each actor, and clarify all possible doubts.

In this first call, companies with the following characteristics were sought to integrate the LEEN:

- Companies of similar size/energy intensity and with considerable energy consumption.
- Companies from different sectors avoid mistrust among their participants and open more easily to share information and exchange experiences.
- Companies located in the same region –within a circle of a maximum of 100 km in diameter–to facilitate the exchanges between its participants.
- Companies that already had an ISO-based management system such as ISO 9001 or ISO 14001:2004 implemented.

- Personnel from companies with experience in energy consulting issues.

More than 60 different companies attended this workshop, including consultants specializing in energy efficiency and interested certification organizations, among others. During this event, companies were invited to express their interest in participating in the LEEN-EMS by filling out a technical sheet in which they described relevant characteristics of their company.

Also, GIZ held a workshop on EMS and LEEN led by a specialized German company with duration of five days. The workshop had the participation of 30 people, including staff from CONUEE, GIZ, consultants, energy efficiency specialists, possible initiators of other LEEN, and other international cooperation organizations.

As a result, 12 companies that expressed interest and presented their general information were selected for a pilot project, based mainly on their energy use intensity and potential to implement energy efficiency measures.

In this arrangement, CONUEE assumed the role of initiator of the LEEN and, jointly with the GIZ, the responsibility of providing support to the LEEN. GIZ also assumed the financing, which included, in turn, the preparation of the energy diagnostics (ED) and the contract of moderation and technical support. The participating companies, for their part, defined a plant in which the EMS would be implemented within the framework of the LEEN. It appointed two responsible people (for each company), who would participate actively and continuously in the activities of the LEEN. It shares relevant technical information for the realization of the ED and commits to hosting at least one LEEN workshop, in addition to covering their related expenses with its exchange activities.

In a survey done with participants for the evaluation at the end of the pilot project, several aspects of the activities were highlighted, but two were identified as very relevant by the participants (GIZ, 2016):

- **Visits to industrial facilities**: In general, the participants mentioned that the visits to the facilities allowed them to learn about new processes, learn from the experiences of other companies, and learn about new focused activities for the optimization of energy consumption. Specifically, the tours generated the greatest exchange of experiences, since they received advice, observations, and congratulations during these tours. In addition, those who visited had the opportunity to see how certain energy efficiency measures or ESM communication campaigns are implemented. This is a source of ideas to implement in their companies.

- **The exchange of experiences**: This is considered one of the pillars of the LEEN, and the participants recognized it as one of its main benefits, as it helps them reduce EMS implementation times, learn from the experiences of other companies, share knowledge, and contact people who are experts in different topics and willing to help each other.

The 12 participating companies of this first LEEN identified opportunities for 25 projects to be implemented in the coming years, in areas such as lighting, cogeneration, compressed air, refrigeration, and sensitization, among others, which are estimated to represent a saving of 27.35 GWh per year, equivalent to 1.75% of the baseline defined for the set of companies.

It is important to note that these activities were accompanied by robust documentation, testimonial videos, manuals, and tools that later provided continuity to PRONASGEn's work.

With support in principle from international allies such as GIZ, ADE, CCA, PTB, and AMEXCID; and with local allies such as the Mexican-German Chamber of Commerce and Industry (CAMEXA), the Mexican Association of Energy Management Companies (AMEXGEN), and the Mexican Association of Energy Efficiency Companies (AMENEER), the activity of the LEEN evolved from networks with financial support from international cooperation to self-financed. Six LEEN-type processes were carried out with international support in a variety of industrial facilities in the private and public sectors, as well as in buildings, under a variety of modalities.

Also, given the success in México, the practice of EMS was promoted under the LEEN scheme in Argentina, El Salvador, and Nicaragua under the initiative of GIZ and support from AMEXID and CONUEE leadership.

At the end of 2018, 17 LEENs had been implemented under PRONASGEn with 169 participating companies or organizations[2] (Table 3.1).

---

[2]PROGRAMA NACIONAL PARA SISTEMAS DE GESTIÓN DE LA ENERGÍA 2013–2018. https://www.conuee.gob.mx/transparencia/boletines/INFORMES/SGEs/Informe_PRONASGEn_Final_30042019.pdf.

Table 3.1 Development of the LEEN in PRONASGEn

Year	Learning Network	Cooperation	Number of companies participating
2015	Industry	GIZ	11
2015	EMS at oil refineries	ADE	2
2015	EMS at dairy and food industry	ADE	4
2015	Pilot project: Introduction of energy efficiency and EMS in SMEs	PTB	21
2015	Public buildings	GIZ	10
2016	Pilot program of North America	CCA	3
2016	Industry	Self-financing	8
2016	PCTr El Salvador	GIZ	11
2017	PCTr Nicaragua	GIZ	9
2017	Bosch	CAMEXA/GIZ self-financing	11
2017	Energy efficiency in electric motors	ICA Procobre	18
2017	Car industry cluster of Nuevo León	CAMEXA/GIZ Self-financing	8
2018	PCTr Argentina	GIZ	12
2018	Municipalities	GIZ (Operated by CAMEXA/AMENEER/AMEXGEN)	9
2018	Transport	GIZ (Operated by CAMEXA/AMENEER/AMEXGEN)	9
2018	Commercial buildings	GIZ (Operated by CAMEXA/AMENEER/AMEXGEN)	12
2018	Industrial parks	GIZ (Operated by CAMEXA/AMENEER/AMEXGEN)	11

Source: Own elaboration.

In total, close to 1,000 professionals related to the EMS process and more than 4,000 indirectly were trained under different arrangements of the LEEN strategy.

By the end of 2018, many of the facilities that had participated in PRONASGEn received technical support from different international organizations and/or participated in the different LEENs that were formed, which had important results in identifying energy efficiency opportunities and improving their energy use.

- In Bimbo Puebla, 25% electricity and 37% natural gas savings were achieved, with an amortization of 2 years.
- In Barilla, natural gas savings of 20% and 13% more capacity were achieved in the set-up of its boilers, with 12% savings in the control of the cooling plant.
- Alpura achieved savings in natural gas of 55% and 46% in electricity by upgrading its pasteurizer.
- McCormick increased its capacity by 9% and obtained electricity savings of 25% by controlling the cooling plant.
- In the food industry in Mexico, an energy saving potential of 25% was located by implementing energy efficiency projects within the framework of an EMS, with payback periods of less than 4 years.
- In PEMEX, the identified actions could achieve savings of 294,810 GJ/year, equivalent to 2,144 million dollars per year in Tula, and 201,481 GJ/year, equivalent to 7.25 million dollars per year in Minatitlán.
- At Bosch, more than 60 energy efficiency and improvement projects were identified.
- Companies participating in the LEEN of electric motors replaced 2,209 motors within two years, where 1,251 motors are from 1 to 20 HP and 958 from 20 HP onwards, with an annual economic saving equivalent to 73 million Mexican pesos.

Also, the promotion process in Mexico has led to interest in ISO 50001's certification:

- Four companies/organizations obtained certifications – Ingersoll Rand Monterrey, Sherwin Williams in El Salvador, a PEMEX Refinery, and the Tecnológico Nacional de México.

- Six companies of the first LEEN (which had continuity under the self-financed modality) expressed interest in seeking certification in ISO 50001 in 2019.

3.1.4 Summary, Conclusions, and Lessons Learned

Based on the need to generate a common work platform with international cooperation, CONUEE took the initiative to design and implement PRONASGEn as a program of national scope aimed at supporting and coordinating activities to promote EMS.

The promotion of EMS in accordance with ISO 50001 was based, mainly, on the promotion, creation, and development of LEENs. In general, the activities to promote the practice of EMS were carried out in work processes with groups of professionals, either from a single organization or several together, always guided by technical support.

Under a very flexible strategy and CONUEE's leadership, working groups and LEENs were promoted and created in various sectors and areas such as the public (federal and municipal), private (large companies and SMEs), and various energy-use sectors (industry, buildings, transport, and municipal services). This process involved:

- Resources from seven organizations, namely, CONUEE, GIZ, PTB, ADE, CCA, DOE, and AMEXCID.
- The creation of 21 learning networks and/or working groups.
- The participation of 167 public and private companies, and/or entities with 254 facilities.
- The direct training of almost 1000 professionals related to the process of the EMS, and more than 4000 indirectly.
- The participation of various public and/or private organizations related to the efficient use of energy and/or processes related to EMS, such as ANCE, CAMEXA, AMEXGEN, AMENEER, universities, and institutes.

It is important to note that these activities were accompanied by robust documentation, testimonial videos, manuals, and tools that were both developed and used by the participants.

This experience demonstrated that the LEEN methodology is a very appropriate and successful mechanism to introduce organizations to EMS and its application through ISO 50001, for their character of collaborative space, which can be successfully applied not only in industrial facilities but also in facilities in the service sector (such as office buildings and stores), transport companies, and municipal service facilities.

There are two elements of the LEEN methodology that, according to the participants, are very relevant: **Visits to the plants of the participants and The exchange of experiences** [3]

Finally, the actions of PRONASGEn and the implementation of LEENs for EMS allowed Mexico to be positioned at the forefront among the countries of the Group of 20, the IEA and IPEEC, for the adaptation and implementation of behavioral change measures.

An EMS is a practice that has been structured in an international standard, ISO 50001, with the perspective that the efficient use of energy should not be an isolated work, but obey a programmed process and with the intervention of all those who participate, directly and indirectly, in decisions related to energy.

Case studies have demonstrated the value of implementing EMS, with energy performance improvements of at least 10% or more of the facilities or systems where they are applied, with most energy savings from low (or no-cost) operational improvements.

Some of the main elements of EMS practice have been part of CONUEE's programs since its origins more than 20 years ago as the National Commission for Energy Saving (CONAE), particularly those aimed at the APF and PEMEX.

However, it was not until the entry of a new federal administration in 2013 that the recent publication of ISO 50001 was identified as an opportunity to intensify the promotion of EMS, not only in the APF itself and in PEMEX, but also in the private sector companies, so that the CONUEE includes related actions in its Annual Work Plan (PAT).

Some of the main lessons of this process are listed below.

Firstly, and extremely important, the EMS is a fundamental practice whose benefits justify their cost. This was demonstrated in the results obtained in most of the facilities that have participated in PRONASGEn above mentioned.

---

[3]GIZ, 2016, DOCUMENTO MEMORIA DE LA RED DE APRENDIZAJE PARA LA IMPLEMENTACIÓN DE SISTEMAS DE GESTIÓN DE LA ENERGÍA EN LA INDUSTRIA. https://www.giz.de/en/downloads/giz2016-es-Documento_Memoria_de_la_Red_de_Aprendizaje_SGEn_Industria.pdf.

It is important to note that in 20 SMEs, savings of more than 57.7 GWh/year were achieved, equivalent to 62 million Mexican pesos, with improvements that have not only managed to reduce energy consumption, but also allowed improvements in the use of their other management systems, personnel, and even obtained improvements in product quality.

Secondly, the processes associated with the EMS allow identifying opportunities for energy and process improvement of companies, some without investment cost and others with sufficiently short periods of return on investments. In the 12 participating companies of the first LEEN in industries, in the coming years 25 energy efficiency projects will be carried out in areas such as lighting, cogeneration, compressed air, refrigeration, and sensitization, among others. These projects are estimated to represent a saving of 27.35 GWh per year, equivalent to 1.75% of the baseline.

Thirdly, a LEEN is a very useful and successful tool for a robust implementation of an EMS, in particular, because of its collaborative space. The fact that today the model is reproduced in a self-financed way, due to the interest of diverse companies and organizations, is a clear sign of its success.

Fourthly, the promotion process in Mexico has led to interest in ISO 50001 certification. PRONASGEn has been, in particular, a large and successful training and dissemination program in best practices for energy efficiency, far exceeding its original global goal and scope. The activities of the LEEN allowed more than 3,000 people to be trained directly or indirectly, and others were sensitized on energy efficiency and EMS.

It is very important to note that for the work of PRONASGEn additional resources were not sought or used, but functions were reorganized within the CONUEE to operate the program in a transversal manner and complement existing programs, such as users with high consumption patterns (UPAC), the Federal Public Administration (APF), states and municipalities, and medium and small enterprises.

Sixthly, the LEEN methodology can be successfully applied not only in industrial facilities, but also in the service sector (such as office buildings and shops), transport companies, and service facilities.

In turn, the success of these activities implies new challenges, particularly those associated with the ability to serve a growing number of companies and institutions under acceptable quality parameters and within the collaborative spirit that has characterized PRONASGEn.

Finally, it is very important to highlight that PRONASGEn has been a success story in terms of international cooperation for Mexico. In the implementation of its various actions, government agencies and institutions from different countries of America and Europe, as well as organizations in the multilateral sphere, converge in a framework of close collaboration and effective coordination, where the CONUEE attends the highest-level conversations on best practices in the field. In the same way, it has served as a reference for a growing portfolio of cooperation offered to third countries by the Government of Mexico, particularly toward Central and South America.

This concept establishes a practice that allows, through the modification of innovation behaviors and access to cutting-edge technologies, to enhance the effects that the different actions of energy efficiency have in specific sectors.

3.2 Experience with Learning Networks in EMS at the Municipal Level[4]

3.2.1 Brief Context of the Project

In Mexico, the energy transition has many areas of development opportunity at the state and municipal levels, especially considering that municipalities under financial pressure could benefit from using local resources for renewable energy and reducing costs through energy efficiency measures.

States and municipalities must be more fully integrated in the implementation to anchor the energy transition at all levels of public administration at the national level.

---

[4]Further information on these municipal networks can be found here: for the Tabasco network: https://www.bivica.org/file/view/id/6272; for the Yucatán: https://www.bivica.org/file/view/id/6273; for Oaxaca: https://www.bivica.org/file/view/id/6274, Energy Efficiency Pumping Network: https://www.bivica.org/file/view/id/6271.

One of the main challenges facing work at the subnational level is the diversity and number of municipalities in Mexico, especially in terms of technical and organizational planning capacities in issues related to energy and its performance.

In this context, it was identified that working under Learning Networks can be a great ally to work simultaneously with several states and municipalities, achieving through collective action and the exchange of experiences to advance more.

The project of Learning Networks in Energy Management Systems at the municipal level was implemented within the framework of the program *"Support for the Implementation of the Energy Transition in Mexico (TrEM) of the German Cooperation for Sustainable Development of Mexico,"* GIZ. Table 3.2 describes the municipalities that participated during the process.

Table 3.2 Result of the selection of the states and their municipalities participating in the learning working of EMS in municipalities

State	Municipalities
Tabasco	Centro, Nacajuca, Cárdenas, Huimanguillo, Jalpa de Méndez
Yucatán	Mérida, Conkal, Progreso, Tecoh, Tizimín, JAPAY
Oaxaca	San Sebastián Tutla, Santa María Atzompa, Santa Cruz Xoxocotlán, Villa de Zaachila, Oaxaca de Juárez, Culiápam de Guerrero

Source: Own elaboration.

One of its main objectives is to develop cooperation to strengthen capacities in countries where it has international cooperation programs. Therefore, being able to "export" tools, instruments, methodologies, and technical and digital innovation to different countries is undoubtedly a great help in bringing the gap in less developed countries and supporting the 2030 Agenda to leave no one behind.

Without the participation, institutional structure, and commitment of the highest local authorities, the implementation of municipal networks is not possible. In addition to being the main beneficiaries, they must be the implementers of the different actions because one of the main objectives of the networks is the strengthening of capacities through technical accompaniment.

In addition, without commitment from the beginning and political will, the identified energy efficiency measures based on energy diagnostics will hardly be implemented.

3.2.2 Barriers to the Implementation of Energy Efficiency Actions in the Municipalities of Mexico

The main barriers faced by municipalities in Mexico can generally be identified as follows:

- Capacity building
- Economic and human resources to improve energy performance
- Social and political decomposition
- Many of them face a reality from poverty to extreme poverty
- Non-existent environmental education programs and actions

3.2.3 Main Results

Carry out energy diagnostics in the 17 municipalities. For the first time, this allowed us to identify areas for improvement and opportunities to make energy efficient, save energy and economic resources, and bring environmental benefits.

- More than 103 people were involved in the activities of the networks.
- More than 8 technology centers and universities from the states participated.

The project had many positive externalities and unplanned positive impacts.

It was possible to involve different universities and technology centers, from each of the participating states, with the aim that young students could participate in the implementation of the networks, and above all strengthen their technical capacities, accompanying the realization of energy diagnostics.

This activity detonated in that, at the end of the project, at least two of the students were hired directly by municipalities as energy managers.

In addition, having resources due to the hybrid implementation of the network, it was possible to implement an additional network of energy efficiency in pumping in the three states.

Other positive impacts were:

- Selected projects that could be financed with their resources and projects for which external financing is required.
- The characteristics and requirements of different sources of financing were analyzed, such as Banobras, Conagua, and different commercial banks.
- Executive projects were developed to accompany the request for financing.
- Institutional capacities were strengthened in energy efficiency, project formulation and evaluation, and financing management.
- The first steps were taken for the operating agencies to implement some energy efficiency measures identified in the short term.

3.2.4 Main Challenges Identified and Overcome

At the beginning of the implementation of the project, the COVID-19 pandemic was declared worldwide, and confinement was taken as a sanitary measure. This confronted us with designing a way to implement the Learning Network in a hybrid way, where most of the workshops were conducted virtually, and the energy diagnoses were carried out in the field with technical experts.

As soon as possible, work began on face-to-face workshops, already toward the closure of the project. Even though networks like this had not been implemented, where the face-to-face part is fundamental for the exchange of experiences and establishing bonds of trust, among other things, exceptional results were achieved, even with the circumstances that were lived worldwide.

- **Meteorological events:** These events such as hurricanes caused personnel of municipalities and the state to pause their participation to attend to the urgency due to floods, etc.
- **Commitment of municipal authorities**: One of the main challenges is to convince the staff of the municipalities of

the importance and benefits of improving their energy management systems (public lighting, buildings, and drinking water pumping networks).

In general, most municipalities in Mexico have limited human and economic resources, so engaging in a process such as the Learning Network, which has a minimum duration of one year, is highly complicated for them. In this case, the role of the state governments as "initiators" of the networks was fundamental, with this it was possible to have permanent accompaniment in each of the 17 municipalities, to achieve attendance at the workshops, accompany the energy diagnoses, and achieve in some cases the beginning of the implementation of identified energy efficiency measures.

Another challenge is the lack of financing and/or incentives for the implementation of energy efficiency measures to improve their energy management systems. Given this, it was possible to implement different measures, first, a mapping of incentives was carried out to implement energy and climate change measures nationally and internationally. It was accompanied by the design and implementation of a course on the elaboration of successful proposals, in which municipalities and states participated in choosing, writing, and developing a request for financing for any of the measures identified. The result was that nine proposals were finalized with the necessary technical support and were ready for the request for resources.

3.2.5 Final Reflections, Lessons Learned, and Next Steps

One way to continue implementing networks would be to involve different energy agencies at the state level, which together could support different initiatives in their respective states.

Collaboration with different actors is an indispensable success factor for implementing projects and initiatives, especially in collective action projects such as Learning Networks.

It was shown that implementing networks in a hybrid way is possible, which considerably lowers the costs of the project, achieves highly effective results, and promotes an environment of trust and commitment among the participants.

The experience of developing three states of the learning networks in energy management systems in municipalities in the context of the pandemic and administrative changes due to the municipal elections also generated good results from the methodological point of view.

Finally, we worked with digital tools to develop a new concept of learning networks in energy management systems: a hybrid network, that is, partly with face-to-face activities and partly with virtual activities.

Chapter 4

The Argentinean Experience on Learning Energy Efficiency Networks

Claudio Carpio[a] and Marina Yesica Recalde[a,b]
[a]*Fundación Bariloche, Argentina*
[b]*Consejo Nacional de Investigaciones Científicas y Técnicas (CONICET), Argentina*
claucarpio@gmail.com, mrecalde@fundacionbariloche.org.ar

4.1 Energy Background of the Country

The Argentinean Energy Mix

The Argentinean energy sector is mainly dependent on hydrocarbons, which in 2022 accounted for 85% of the total primary energy supply, with natural gas as the most relevant fuel (53% out of TPES). As can be seen in (Figure 4.1), the role of natural gas has increased significantly from t sixties and seventies, especially since the discovery of the field of Loma La Lata in 1977 and as a result of a clear public energy policy oriented to the substitution of coal and oil by natural gas, and the penetration of this fuel in electricity and

Learning Energy Efficiency Networks: Applications and Challenges in Latin American Countries
Edited by Claudio Carpio and Marina Yesica Recalde
Copyright © 2026 Jenny Stanford Publishing Pte. Ltd.
ISBN 978-981-4968-82-9 (Hardcover), 978-1-003-77468-6 (eBook)
www.jennystanford.com

DOI: 10.1201/9781003774686-5

many final consumption sectors (Recalde, 2011). This role seems to continue as the country has significant reserves and resources of this fuel.

Despite recent efforts made to increase the role share of renewable energies in the power sector, in 2022 55% of energy consumption in power generation was covered by natural gas. As argued in Recalde (2011b), the relevance of natural gas in power generation increased during the nineties. Perhaps, the main reasons have been the lower cost of thermal power generation in relation to hydroelectric one; changes in the policies of multilateral organisms of credit; improvements in natural gas transport and distribution; lower environmental impact of natural gas in comparison to fuel oil; and technological innovations in combined cycle natural gas plants (Guzowski and Recalde, 2008; Guzowski and Recalde, 2008b). Moreover, some authors argued that these investments have been strongly associated with the local availability of natural gas at low prices during the nineties and the strategies of private producers (Pistonesi, 2000; Pistonesi, 2001).

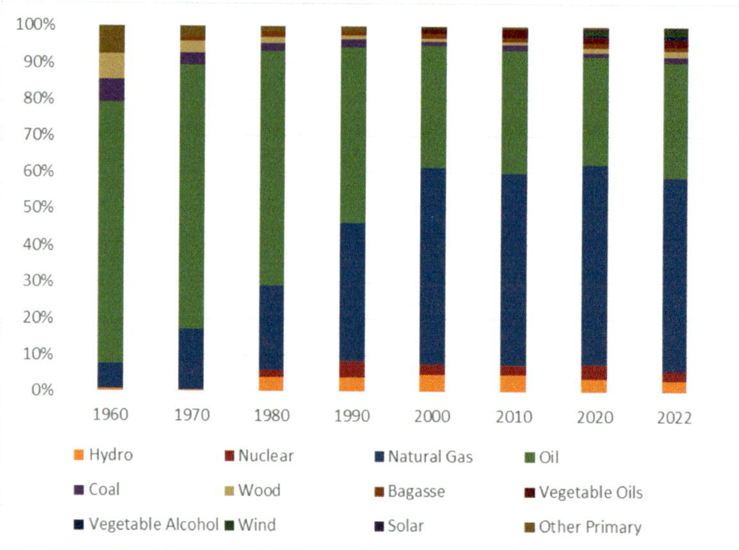

Figure 4.1 Total primary energy supply evolution 1960–2022.

Source: Author's analysis based on National Energy Balances of the Secretary of Energy.

4.1.1 The Evolution of Sectorial Energy Consumption

In 2022, the industrial sector accounted for 23% of the total energy consumption (including primary and secondary sources) (Figure 4.2). This share of the industrial sector within domestic final energy consumption has been reduced from 36% in 1970 (Figure 4.3). The role of industrial energy consumption was around 30% until the end of the 2000s. From 2010 onwards, its relative relevance was reduced, while the role of households and transport sectors increased.

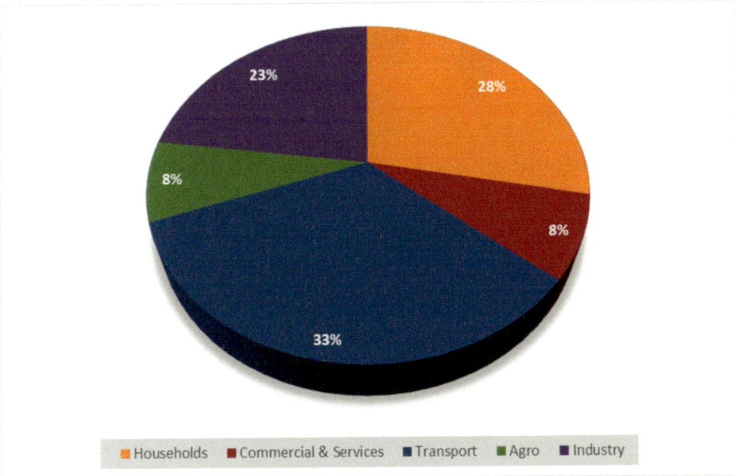

Figure 4.2 Total energy consumption by sectors in Argentina (2022).

Source: Author's analysis based on National Energy Balances of the Secretary of Energy.

Argentina's national energy balance is not disaggregated in subsector's consumption which makes it difficult to evaluate the implementation of the most relevant industrial sectors to develop energy efficiency strategies to mitigate the energy impact of consumption. To this purpose, in the context of the EU-financed project Energy Efficiency in Argentina, researchers and experts developed a specific study to estimate the sub-sectorial energy consumption within the industrial sector (Dubrovsky et al., 2020). As shown in Figure 4.4, the authors estimated that iron and steel, petrochemicals, oil, cement, aluminum, and some other foodstuffs are that among the most energy-intensive sectors. There is also a set of very relevant consumptions, associated with the consumption of SMEs, which are very important.

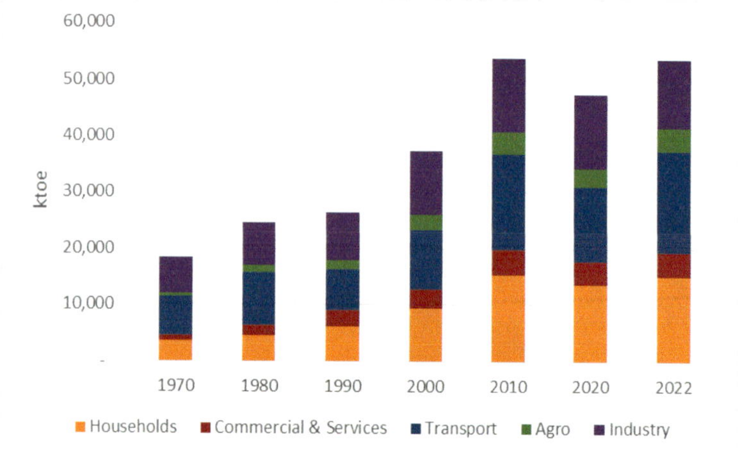

Figure 4.3 Evolution of total energy consumption by sectors in Argentina (1970–2022).

Source: Author's analysis based on National Energy Balances of the Secretary of Energy.

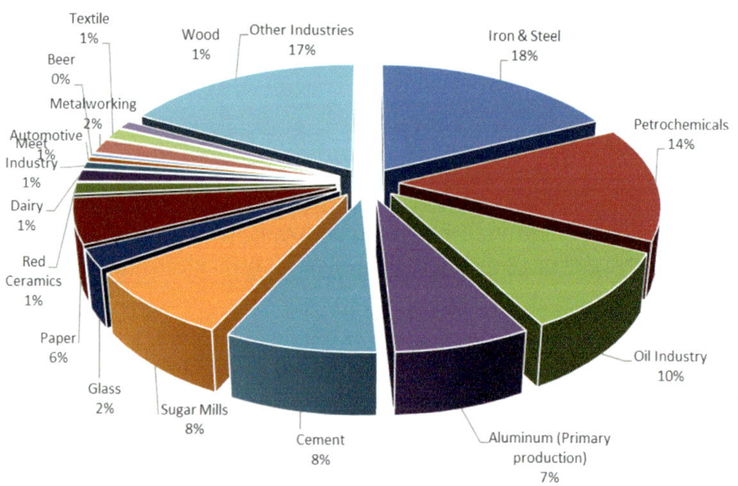

Figure 4.4 Industrial energy consumption by activity (2019).

Source: Dubrovsky et al. (2020).

The industrial sector is mainly dependent on natural gas, which in 2022, represented 61% of its energy consumption, followed by electricity (31%), which has been the result of fuel substitution, as can be seen in the historical evolution shown in Figure 4.5.

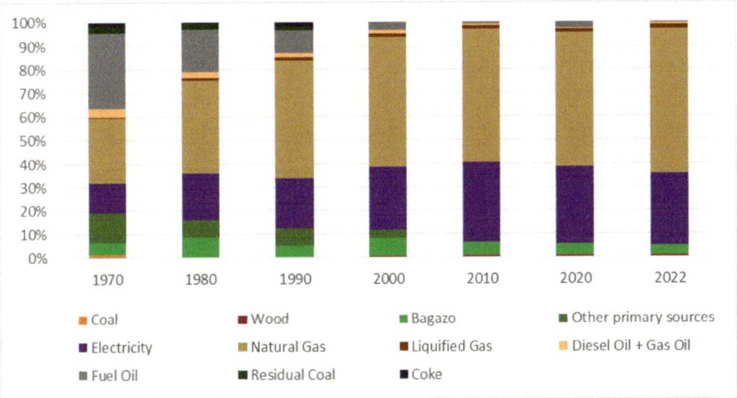

Figure 4.5 Evolution of total energy consumption in the industrial sector by fuel (1970–2022).

Source: Author's analysis based on National Energy Balances of the Secretary of Energy.

4.2 Energy Efficiency Policies and Activities in Argentina during 1980–2021

4.2.1 The 1980–1990 Period

Argentina initiated its journey by the energy efficiency pathway at least five decades ago. The first action to institutionalize the rational and efficient use of energy was, in late 1979, the creation of the National Directorate of Conservation and New Energy Sources, under the authority of the Energy Secretariat.

On of the first and most ambitious program for renewable energy consumption and energy efficiency during this period was the National Program for Rational Use (Decree 2,247/85), inspired by many of the policies implemented in developed countries since 1975 (Calleja, 2005: 121). This program was composed of three subprograms: the energy conservation subprogram; the fuel substitution subprogram; and the subprogram for the evaluation, development, and application of new energy sources. The energy conservation subprogram aimed at improving energy efficiency in all consumer sectors including different activities, from fuel substitution processes to agreements and dissemination campaigns for education and research on energy and environmental topics (Bouille, 1999; Camarda, 2019). The Energy Substitution Program sought to replace fuels (replace scarce liquid fuels, with more abundant ones, such as natural gas or renewable sources such as biomass ethanol; and replace imported fuels with domestic ones). In the case of the new energy sources plan, some of the most relevant objectives were to provide energy to deprived regions of the country through the application of proven technologies; replace, where feasible and convenient, the use of fossil fuels with unconventional energy sources; and improve the recovery of fossil fuels from depleted deposits.

In 1985, the Regional Wind Energy Center (CREE) was created, and in the same year, one of the first estimates of the national wind energy potential was made. In 1987, the National Energy Secretariat published a "National Inventory of Small Hydroelectric Developments," which covers up to 5 MW of power (Recalde et al., 2015).

The main actions implemented in this context include the creation of the Energy Studies Groups (GESE as it stands in Spanish), the National Technological University (UTN), the agreements with the French Agency for Energy Management (AFME, France) and the Institute for Diversification and Energy Savings (IDAE, Spain), and the Rational Use of Energy training program in schools in Buenos Aires, Rosario, and others in the province of Santa Fe.

4.2.2 The 1990–2000 Period

This decade was characterized by a changing macroeconomic context with the State Reform Law (Law 23,696) and the Economic Emergency Law (Law 23,697) being the key instruments for the development of many of the policies implemented in the energy sector, and the foundations for the privatization of the main state-owned companies.

Different authors argued that market deregulation and privatization of companies, the introduction of private rationality, the vertical disintegration of the energy chains, the lack of control, and the increase in the extraction and exploitation of primary resources resulted in a depletion of the system and a lack of sustainability that led to the energy problems evidenced since the 2000s (Azpiazu, 2008; Basualdo and Azpiazu, 2002; Kozulj, 2002; Pistonesi, 2001; Recalde, 201). At the same time, in March 1991, the government passed the Austral Convertibility Law, which established a fixed exchange rate between the U.S. currency and the Argentine peso (Argentina). The establishment of this new monetary policy had great repercussions on the economy and the energy sector (Basualdo and Azpiazu, 2002; Pistonesi, 2002).

Bouille (1999) and Recalde et al. (2015) stressed that these aspects were shaping environmental conditions that propitiated a negative framework for rational use of energy policies, resulting in a low level of success in terms of the impact in final consumption sectors. This may be true, even when two of the most relevant laws of this period regulating electricity and natural gas mentioned efficiency aspects. Firstly, Law 24.065/92– Electricity Regime (1992) in its Article 2, item e –encourages the supply, transport, distribution, and efficient use of electricity by setting appropriate tariff methodologies. Secondly, Law 24.076 –Natural Gas, Regulatory Framework (1992), article 2, and item f—encourages the rational use of natural gas and ensures adequate environmental protection.

During the 1992–1999 period, the Program for the Rational Use of Energy (URE RA-UE) was implemented to define technical, economic, and financial instruments that facilitate the efficient use of energy in certain priority areas (households, public lighting, and transport). Some of the most important actions performed during these years were the creation of the National Directorate of Promotion (Rational Use of Energy and Renewable Technologies) and the technical cooperation with the European Union. This Program was constituted of fourteen subprojects: (1) Industrial Sectoral Diagnosis from the Perspective of Rational Energy Use; (2) Sectoral Diagnosis of Transport from the Perspective of Rational Energy Use; (3) Study of the potential of energy cogeneration in the industrial sector of the Province of Buenos Aires—Development of Regulations; (4) Application of technical solutions and technology transfer on electrical energy consumption measurement and monitoring equipment in SMEs: (5) Energy Quality Certification in Appliance Equipment; (6) Technical Assistance to the RUE Unit: Institutional; (7) Creation of URE Database—Website Development Internet; (8) Energy and Light Optimization in Municipal Public Lighting Systems; (9) Actions of Rational Use of Energy in Grain Drying; (10) Energy Cogeneration in the Tertiary Sector; (11) Training Actions in the Industrial and Services Sector; (12) Actions to measure and monitor combustion in diesel engines in urban and intercity freight transport; (13) Optimization of combustion and heating processes in the steelmaking sector; and (14) Optimization of combustion and heating processes in the ceramic sector (Bourges, 2013).

Between 1998 and 1999, the government implemented the Program for Households Devices Quality (PROCAEH for its acronym in Spanish), with the main objective of reducing the overall consumption of electricity by using more efficient appliances, through introducing a mandatory labeling system for the most relevant devices. This program was the seed of subsequent efforts that resulted in the current energy labeling system in household appliances and dwellings in Argentina (Bourges, 2013).

4.2.3 The 1999/2000–2010 Period

These 10 years were very changing in terms of enabling conditions for the macroeconomy, the energy sector and, therefore, energy efficiency policies. National activity fell in 2001 (4%) and the GDP fell by 10% in 2002. Political, economic, and social discontent led to the outbreak of a protest that culminated in December 2001 and ended with the resignation of the President. In January 2002, the "*Law of public emergency and reform of the exchange regime*," Law 25,561, was enacted, which declares a public emergency in social, economic, administrative, financial, and exchange matters. One of the main aspects of the law was the modification of the articles of Law 23,928, which established convertibility, abandoning the fixed exchange rate regime. This change in the macroeconomic environment impacted on the structure of the energy sector by changing the context in which contracts are developed. The prices and tariffs resulting from such clauses were established in pesos at the one-to-one exchange rate with the dollar, locally known as the "pesification" of public utility tariffs. In 2003, the Argentinean economy began to recover. According to data from the National Institute of Statistics and Census (INDEC), between 2003 and 2007, the GDP grew at a rate exceeding 8% per year. In 2004, symptoms of the energy crisis began to show, mainly related to the availability of natural gas, a fundamental resource of the national energy system, because of a growing demand for energy and a reduction in the levels of hydrocarbon production and reserves. In this context, the government began to implement different measures, which, however, were not enough to mitigate the supply problems that would deepen throughout this and the following decade. At the same time, the lack of investments in electricity generation, the restrictions in the electricity and transportation sectors, and the problems of natural gas supply led to a review of the role of the state in the sector (Recalde et al., 2015).

Regarding energy efficiency policies, in the 1999–2005 sub period, many actions were supported by external cooperation. In this case, through the German Technical Cooperation Agency (GTZ), which years later became the GIZ. The Program to Increase Energy and Productive Efficiency in SMEs (PIEEP), aimed to improve the competitive conditions of Argentina's small and medium-sized enterprises, by implementing demonstration units, training workshops, and studies in companies in various sectors, such as grain drying in Santa Fe and Jujuy, dairy products and metalworking in Cordoba, agricultural irrigation, fruit preserves, and ferroalloys in Mendoza, vegetables in Río Negro, citrus and sugar mills in Tucumán, and dairy products and industrial parks in Buenos Aires (Camarda, 2019).

Camarda (2019) argued that Decree 27/2003 had significant impacts in terms of the role of the state. Article 2 of the Decree stated that the Energy Secretariat is responsible for analyzing the dynamic of energy markets and elaborating the strategic planning in electricity, hydrocarbons, and other energy markets, through the promotion of competition and efficiency policies in the resource allocation process; and must also lead the actions aimed at applying the sectoral policy, leading the process of adaptation of the new operators to the general interest, respecting the rational exploitation of resources and the preservation of the environment.

In 2003, the Government launched the Public Buildings Energy Savings and Efficiency Program (PAyEEP for its acronym in Spanish) to reduce overall energy consumption in National Public Administration buildings. This program was also connected to the Renewable Energy for Rural Markets (PERMER for its acronym in Spanish) program. These resources were used to finance four pilot experiences that were carried out in public buildings in the cities of Neuquén, Jujuy, Tucumán, and Buenos Aires.

Then, in 2004, the government implemented two rational uses of energy-oriented plans. The Rational Energy Use Program (PURE for its acronym in Spanish) (Resolution SE 415/2004) to encourage the reduction of consumption (both natural gas and electricity) of all residential and commercial users in relation to their consumption of the same periods in 2003, using a credits and charges system. Camargo (2019) argued that a weakness of these programs was that they did not include energy supply and the energy value chain.

To include other consumers not incorporated in these programs, Resolution 552/2004 created a program for the three more relevant power utilities of the country (regarding coverage) (EDENOR S.A., EDESUR S.A., and EDELAP S.A.) including all users, except for public lighting. In a similar way, Resolution SE 931/2005, aimed to implement an Electricity Rational Use Program for Large Users Entering the Wholesale Electricity Market (MEM for its acronym in Spanish).

Recalde and Guzowski (2012) demonstrated that these programs were not successful in achieving the main objectives. According to the authors, the lack of appropriate publicity for the program can be one of the main reasons for this failure. However, the main problem seems to have been energy prices, which remained practically frozen during the period; therefore, consumers did not face enough incentives to reduce their electricity demand. Authors argued that reductions in electricity consumption between 2008 and 2009, particularly in the industrial and commercial sectors have been related to reductions in economic activity, as well as to substitutions within the energy mix. Moreover, from 2006 onward, some industries decided to invest in their power generation, to avoid electricity supply shortages and the consequent economic and financial impacts.

An energy efficiency milestone was the Program for the Rational and Efficient Use of Energy (PRONUREE for its acronym in Spanish) (Decree 140/2007), which states that energy efficiency is understood as the adaptation of energy production, transportation, distribution, storage, and consumption systems, aimed at achieving the greatest sustainable development with the existing technologies, minimizing the impact on the environment, optimizing energy conservation, and reducing costs. In this Decree, Article No. 1 declares the rational and efficient use of energy as of national interest and priority, while Article No. 2 approves the guidelines of the National Program for the Rational and Efficient Use of Energy (PRONUREE for its acronym in Spanish), to improve energy efficiency in all energy-consuming sectors. As listed by Bourges (2013), some of the most relevant activities implemented in the framework of the Decree were: (a) The Program to increase efficiency in the public lighting systems; (b) Labeling standards and Minimum Energy Performance Standards (MEPS); (c) Standard and online application to promote energy efficiency in housing construction (IRAM 11.900), (d) Pilot experiences in agricultural irrigation, textile industries, etc.; (e) Regulation of cooling temperature in public buildings, efficient use of lighting, etc.; (f) National Public Administration building facilities database; and (g) Energy managers.

In 2008, the government passed Law 26.473, and in 2010, the Decree 2060/2010, forbade the import and commercialization of incandescent lamps, accompanied by other instruments such as the free distribution of 25 million compact fluorescent lamps across the national territory.

4.2.4 From 2010 Onwards

In 2009, in a framework of cooperation with the Global Environment Facility (GEF), the country implemented the Energy Efficiency Project in Argentina (2009–2015), which had two main components: (i) Argentine Energy Efficiency Fund and (ii) Enhancing energy efficiency and management capacities. This project, one of the most relevant and close backgrounds for the LEENs, included Energy diagnostics in companies; the Creation of the Argentine Energy Efficiency Fund; Energy Efficiency Regulation; and Normalization and labeling and Development of Energy Service Providers (Energy Service Companies).

The main lines of action linked to this initiative included the Large-Electro Intensive Users Resolution (1E-2017), which granted subsidies on the electric energy bill to priority electro-intensive sectors (23% of the total industrial demand for electrical energy).

On the other hand, a line of credits granted by the Foreign Trade and Investment Bank (BICE for its acronym in Spanish) was in force to facilitate investments in energy efficiency projects, with priority in the SMEs.

In the transport sector, the manufacturing and marketing of new vehicles with low emissions and high-efficiency standards were promoted. In this sense, the first mandatory vehicle energy efficiency label was published. An intelligent Transport Program was carried out, which linked carriers, cargo givers, service and technology providers to adopt strategies, tools, technologies, and knowledge that contributed to improving fuel efficiency in heavy vehicles. It sought to achieve fuel savings of up to 30%. The third axis of work was the promotion of good driving habits, which could help obtain up to 20% fuel savings with very low investment. Regarding the energy efficiency labeling, it was calculated that the projected savings accumulated to 2030 through this policy were equivalent to 7,800 ktoe.

Currently, there is a mandatory label with MEPS for refrigerators and freezers, washing machines, air conditioners, gas cooking appliances, heaters, convection heaters, gas hot water tanks; mandatory label without minimum standard for halogen and fluorescent lamps, ballasts for fluorescent lamps, three-phase and single-phase induction motors, televisions, electric hot water tanks, and microwaves; and voluntary label for household electric pumps, electric ovens, fans, dishwashers, LED lamps, boilers, and windows in houses.

The National Housing Labeling Program aimed to implement a Households Energy Efficiency Label to provide information about the energy performance of a home. For its implementation, a computer application was developed, and pilot tests were carried out in various bioclimatic areas of the country to collect information and define ranges of the energy efficiency scale according to each zone. This initiative included the definition of standards for social housing. From the improvement in the efficiency standards in buildings, a potential decrease in the energy requirement of 50% was detected.

As a result of this program, there are draft provincial laws on housing labeling in Santa Fe, Mendoza, Río Negro, Tucumán, and Salta, and the city of Buenos Aires (CABA). The Program of Rational and Efficient Use of Energy (PROUREE) in Public Buildings aimed to reduce the levels of consumption in the National Public Administration and it assumes an exemplary function before the rest of society. It consisted of implementing measures to improve energy efficiency, managing energy, and raising awareness among staff in the rational use of resources.

The Efficient Lighting Plan consisted of the replacement of luminaries with more efficient equipment of LED technology on public roads, both in municipalities and provincial routes of the country through the transfer of non-refundable funds. As of December 2018, 65 projects for the replacement of public lighting luminaries were fully executed in localities corresponding to 17 provinces of the country, acquiring 86,000 luminaries. This measure allowed obtaining 50% energy saving. Despite having the efficient use of energy objective conveniences for its implementation, its development is limited, among others, by a series of informational and cultural barriers.

For these reasons, it is necessary to contemplate a cultural-educational strategy whose ultimate objective will be the change of habits toward a culture of efficient use of energy. The National Education Strategy for Energy Sustainability responds to this need and aims to develop education programs for all levels of the education system to educate the responsible and efficient use of energy and generate technical capacities in society. The strategy had the approval of the Provincial Ministers of Education throughout the country, meeting in the Assembly of the Federal Council of Education. Content and teaching materials were developed and delivered to all schools in the country; training actions with an impact on more than 75,000 teachers and 130,000 students throughout the country; training cycles in the provincial technical teams and conjunction with the National Institute of Technological Education (INET) and the Secretariat of University Policies; and guidelines were formulated for the inclusion of energy sustainability in Technical Secondary Education, Vocational Technical Training, and the Technical and University Tertiary levels throughout the country.

In 2015, Argentina signed the Paris Agreement, and in 2016, presented its revised National Determined Contributions (NDCs), with the target of not exceeding the 483 net emissions of tons of CO_2eq in 2030. Then it submitted a second in 2020, and in 2021 a revision of the target was submitted, establishing that "Argentina's GHG emissions will not exceed, in 2030, 349 million tons of CO2e". For this purpose, the National Cabinet on Climate Change has put forward different Climate Change Action Plans (one for each of the sectors identified as highly relevant for both mitigation and adaptation actions)[1], in which energy efficiency actions are very relevant (particularly in the Energy, Transport, and Industry National Action Plans). Within these plans, the role of energy efficiency in the industrial sector is a priority.

4.3 Barriers to Energy Efficiency in Industry in Argentina

Bouille et al. (2021) and Carpio and Recalde (2021), based on the main results of the study performed in the framework of the EU-financed project, discussed the barriers to Argentinean's industrial sector. As mentioned by the authors, these barriers were identified through a research process developed over more than two years, which consisted of surveys, interviews, and discussion workshops with key actors in the sector, in the framework of the EU-financed project. This study differentiated between barriers in large, small, and medium companies.

Figure 4.6 shows that the main general barriers identified in the case of large companies were Financing Barriers (lack of specific

---

[1]National Industry and Climate Change Action Plan, National Energy and Climate Change Action Plan, National Agro and Climate Change Action Plan, National Transport and Climate Change Action Plan, National Heath and Climate Change Action Plan. For more information visit: https://www.argentina.gob.ar/ambiente/cambio-climatico/planes-de-accion.

credit lines and the difficulty of access to financing, as well as the high cost of financing); Economic and Market Barriers (low relative profitability of projects, especially replacement projects, high costs of technologies, difficulties of access to imported technologies or their cost); and Technological and Capacity Barriers (related to the lack of technical capabilities for the evaluation of energy efficiency alternatives in the company). There are also some Institutional and Regulatory Barriers (with many references to the lack of regulatory frameworks for cogeneration), Information Barriers, and Cultural and Awareness Barriers (especially in reference to the lack of awareness on the subject at high management levels). Figure 4.7 shows the barriers differentiated by industrial branches, where the preponderance of the different barriers can be seen. In the case of SMEs, the general barriers are similar but with a higher share of barriers related to technological and human capacities and information barriers. In the case of technological and capacity barriers, the problem is related to the internal human capacity to evaluate options and alternatives for energy efficiency actions, financing possibilities, and access to financing (beyond the cost of financing options).

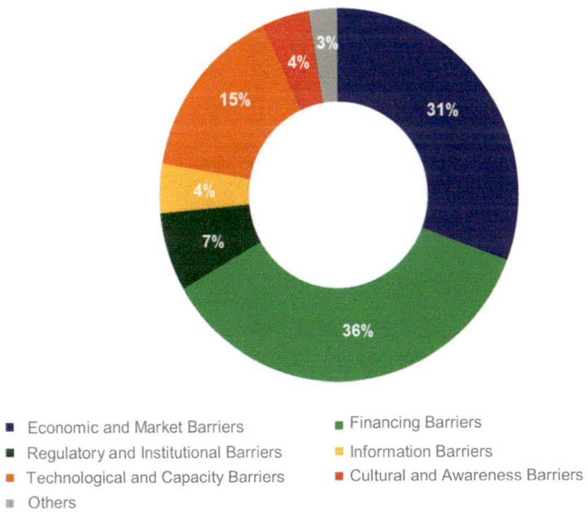

Figure 4.6 Barriers to energy efficiency in large industrial companies in Argentina.

Source: Bouille et al. (2021).

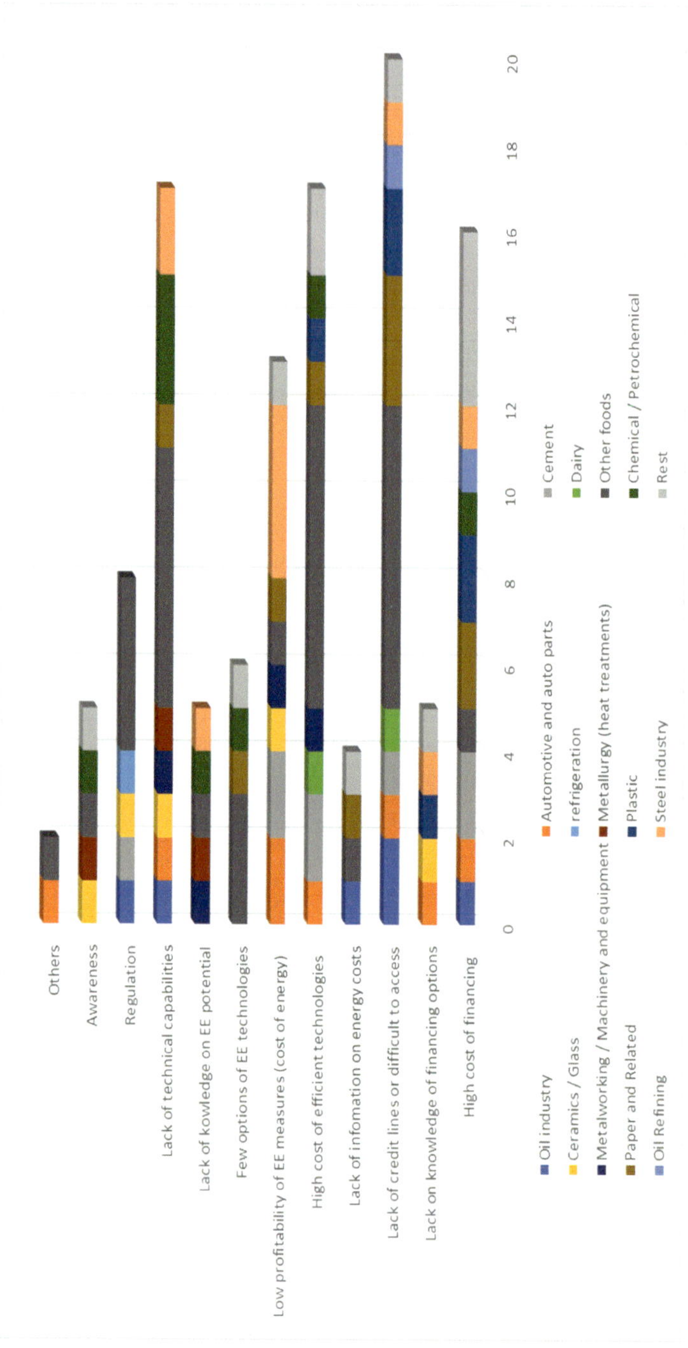

Figure 4.7 Energy efficiency barriers identified in the Argentinean industrial sector.

Source: Bouille et al. (2021).

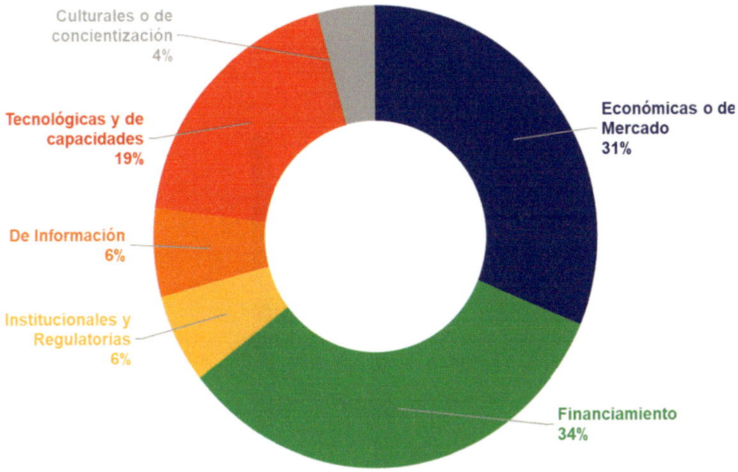

Figure 4.8 Barriers to energy efficiency in SMEs by major categories.

Source: Based on Bouille et al. (2021).

There are also enabling conditions and boundary conditions that have been identified by the key stakeholders involved in the process, such as the impacts of macroeconomic policies and, especially, monetary policy; the need to comply with the Paris Agreement, which impacts positively; and the lack of a well-developed capital market in the country. Some enabling conditions mentioned by different actors or stakeholders include the lack of a long-term energy policy; information problems, for the development of energy efficiency policies and their monitoring and evaluation (M&E), as well as for the evaluation of environmental and climate change policies; lack of relation and coordination within the public and private sector, lack of articulation with the scientific system; R&D&I policy in the industrial sector; problems in the interior of the country to access training and events that distribute knowledge; and lack of predictability in electricity and distributed gas tariffs.

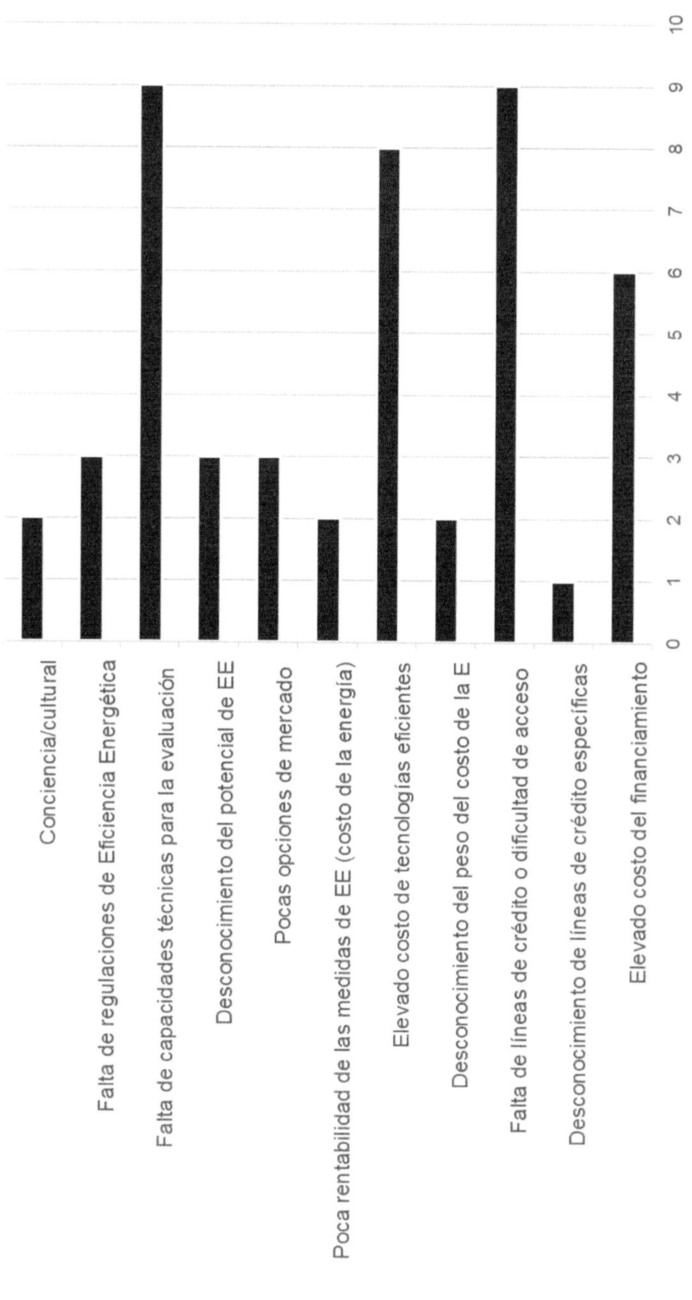

Figure 4.9 Barriers to energy efficiency in SMEs by subcategories.
Source: Based on Bouille et al. (2021).

4.4 The LEENs in Argentina

4.4.1 The First Argentinean LEEN: The Triangular Cooperation Between Mexico-Germany and Argentina

The first LEEN was initiated in 2018 based on a Triangular Cooperation between the existing Argentinean Undersecretariat for Energy Saving and Efficiency, the National Commission for the Efficient Use of Energy (CONUEE) from Mexico, and the GIZ of the German Ministry for Economic Cooperation and Development (BMZ).

This network was implemented in the province of Buenos Aires, more specifically in the Autonomous City of Buenos Aires and Greater Buenos Aires (CABA and GBA). It was made by 12 companies belonging to energy-intensive sectors of the technology, food, paper, iron, steel/metal, automotive, metallurgical, and metal-mechanic industries. The companies that participated in the LEEN were: Acindar Industria Argentina de Aceros SA, BGH SA, Bimbo de Argentina SA, Celulosa Campana SA, Empresa Argentina de Soluciones Satelitales—ARSAT SA, Establecimiento Metalúrgico Sturam SA, Fibercord SA, Molinos Río de la Plata SA, Packall SRL, Ricardo Almar e hijos SA, Roca Argentina SA y, and Volkswagen Argentina SA.

According to the Argentinean government information, the main goals were achieved, with nearly 63% of the implementation of actions related to the EMS, and one company with the aim of certifying the ISO 50001. Table 4.1 shows the energy saved by this first LEEN in relation to its original estimated electricity and natural gas consumption, i.e., 6.8% of electricity and 5% of natural gas.

Table 4.1 Results of GBA-CABA (GIZ) LEEN[2]

	Energy consumption (GWh/year)	LEEN energy avoided (GWh/year)
Electricity	205	14
Natural gas	377	19

Source: Based on information from the Energy Secretariat information.

[2]For more information visit: https://www.argentina.gob.ar/economia/energia/eficiencia-energetica/eficiencia-energetica-en-sectores-productivos/redes-de-aprendizaje-de-eficiencia-energetica.

4.4.2 The EU-Financed Networks: From LEEN 2 to 7

The next six networks of the country were framed by the Energy Efficiency Project in Argentina (EEArg) implemented during 2018–2020, between the European Union and the Argentinean Government (Secretary of Energy). The aim of the project was to support the Government of Argentina in its efforts to modernize its energy sector through the creation of an enabling environment that encourages energy efficiency and renewable energy investments. The project was implemented by a consortium of four international organizations: GFA Consulting Group, Fundación Bariloche, EQO-Nixus, and Fundación CEDDET[3].

One of the key components of the EU-financed project was the promotion of the implementation of the EMS in the Industry, based on the norm ISO 50001, through the implementation of five Learning Energy Efficiency Networks. Therefore, in the specific case of Argentina, the final objective of the LEENs was to promote the implementation of the Energy Management System according to the ISO 50001 standard in the industrial sector.

The first five regional LEENs were created in Santa Fe, Tucumán, Córdoba, Buenos Aires (Industrial Park of Pilar), and Misiones. The five LEENs were established as follows[4]:

1. Santa Fe, 2018
2. Tucumán, 2019
3. Córdoba LEEN, 2019
4. Pilar Industrial Park, 2019
5. Misiones, 2019

These networks are comprised of 60 industrial companies from different sectors of the economy including food, metal-mechanical, automotive, dairy products, textile, chemical, plastics, edible oils, steelmaking, sugar mills, meat processing plants, ceramics, soap, and building materials.

Figure 4.10 shows the location of networks, the number of companies that comprise it, and the industrial facilities involved, as there are cases with more than one industrial facility per company.

---

[3]For more information visit: https://eficienciaenergetica.net.ar.

[4]Date corresponding to the official launch workshop of each LEENs.

Figure 4.10 The first five LEENs implemented in Argentina in the framework of the Energy Efficiency Project financed by the European Union.

Source: Carpio and Recalde (2020).

In 2020, the sixth network was initiated in the wine sector, made up of 22 companies, which was developed entirely in the pandemic context by COVID-19.

The Santa Fe LEEN[5] was initiated in Rosario with 11 companies from different sectors (Figure 4.11), from which 10 finalized the network. During this network, six workshops were carried out, which took place in classrooms and at the facilities of some of the companies.

[5]Based on: https://eficienciaenergetica.net.ar/img_publicaciones/09301931_SANTAFE.pdf.

Figure 4.11 Companies participating in the Santa Fe LEEN.

Source: Argentinean Energy Efficiency Project webpage.

According to the information from the EEArg Project, total energy savings for the LEEN were 2%; 2.3% for electricity and 1.7% for heating, which led to an estimated economic saving of 1.8 MMUSD and 0.9 MMUSD, respectively. One aspect highlighted by the energy experts is that in this LEEN energy, cost accounted for an average of 20.8% of operation costs, which is clearly very relevant to the success of the network.

Among the most relevant barriers identified in this network the organizers highlighted include the lack of trained experts who work for the development of the methodology; the lack of sub-metering in some companies in the network, which will affect the measurement and savings verification plans after implementation; the management (CEO) of the network companies, due to the context (even pre-covid), disconnected causing a bad performance of the network; financial barriers, access financing on affordable terms to undertake medium- to long-term action plans; the lack of correct access to information barrier; and the regulatory barriers, scarcity of public regulations that promote the improvement of energy performance in the industrial sector. As a result of these barriers, one of the companies left the LEEN close to the 3rd Workshop.

In the case of the LEEN implemented in the Province of Tucumán[6], 13 companies were included (Figure 4.12).

Figure 4.12 Companies participating in the Tucumán LEEN.

Source: Argentinean Energy Efficiency Project webpage.

According to the information from the EEArg Project, total energy savings for the LEEN were 3%; 1.7% for electricity and 3.1% for heating, which led to an estimated economic saving of 0.3 MMUSD and 1.3 MMUSD, respectively. The total savings of the LEEN action plan were 1.5 MMUSD. In this case, the energy cost accounted for an

[6]Based on: https://eficienciaenergetica.net.ar/img_publicaciones/09211847_TUCUMAN.pdf.

average of 21.8% of operation costs, which is clearly very relevant to the success of the network.

The third LEEN in Cordoba included 11 companies (Figure 4.13) and was implemented with the collaboration of the Cordoba Energy Committee (CEC), which is part of the Center for Research and Studies on Culture and Society (CIECS), a dual dependency unit integrated by CONICET and the National University of Cordoba.

Figure 4.13 Companies participating in the Cordoba LEEN.

Source: Argentinean Energy Efficiency Project webpage.

In this case, the total energy savings were 3.6% (which represented 0.65 MMUSD), primarily thanks to the savings in electricity (4.8%) and heating (2.1%) (Table 4.2).

In the case of the LEEN implemented in the Industrial Park of Pilar[7], the first event was held at CEPIP in the Pilar Industrial Park, and 14 companies joined the network. In this case, 10 workshops were held; some of them work face-to-face and others virtually (in the context of COVID-19).

[7]Based on: https://eficienciaenergetica.net.ar/img_publicaciones/09272028_PILAR.pdf.

Table 4.2 Overall results of the 11 companies that make up the LEEN-Cordoba

Total energy costs	14.6 MMUSD
Electricity costs	12.7 MMUSD
Heating costs	1.9 MMUSD
% on operating costs	6.9%
Action plan (energy performance improvement)	18 months of participation in the Network
Investment	€0
Total savings (action plan)	0.65 MMUSD
Electricity savings	0.61 MMUSD
Heating savings	0.04 MMUSD
Percentage of global savings	3.6%
Electricity percentage	4.8%
Heating percentage	2.1%

Source: https://eficienciaenergetica.net.ar/industria_redes_de_aprendizaje.php

Energy costs represent 11.5% of total operating costs, and the results were 1.6% (electricity 1.7% and heat 1.5%) (Table 4.3).

Table 4.3 Overall results of the 11 companies that make up the Learning Network—Industrial Park of Pilar

Total energy costs	20.37 MMUSD
Electricity costs	19.61 MMUSD
Heating costs	0.76 MMUSD
% on operating costs	11.5%
Action plan (energy performance improvement)	18 months of participation in the Network
Investment	€0
Total savings (action plan)	0.339 MMUSD
Electricity savings	0.328 MMUSD
Heating savings	0.011 MMUSD
Percentage of global savings	1.6%
Electricity percentage	1.7%
Heating percentage	1.5%

Source: https://eficienciaenergetica.net.ar/industria_redes_de_aprendizaje.php

Figure 4.14 Companies participating in the Industrial Park of Pilar LEEN.
Source: https://eficienciaenergetica.net.ar/industria_redes_de_aprendizaje.php

The Misiones LEEN[8] was special, as it was developed in conjunction with the Argentine Confederation of Small and Medium-sized Enterprises (CAME), to incorporate small and medium-sized enterprises (SMEs) in the experience. The first workshop was held at the headquarters of the Economic Confederation of Misiones (CEM), and initially, 15 companies joined the network (four of them later abandoned the process) (Figure 4.15). As a result of this joint experience with CAME in the LEEN, in 2024, a new project for the implementation of four LEENs for SMEs of Patagonia Argentina, financed by the European Union, will be initiated.

According to the information from the EEArg Project, total energy savings for the LEEN were 4.8%; 12.1% for electricity and 2.7% for heating, which led to an estimated economic saving of 0.29 MMUSD and 0.012 MMUSD, respectively. In this, LEEN energy cost accounted for an average of 13.5% of operation costs (Table 4.4).

[8]Based on: https://eficienciaenergetica.net.ar/img_publicaciones/09272026_MISIONES.pdf.

Figure 4.15 Companies participating in Misiones LEEN.

Source: Argentinean Energy Efficiency Project webpage.

Table 4.4 Overall results of the 11 companies that make up the Learning Network—Misiones

Total energy costs	2,834 MMUSD
Electricity costs	2,397 MMUSD
Heating costs	0.436 MMUSD
% on operating costs	13.5%
Action plan (energy performance improvement)	18 months of participation in the Network
Investment	€0
Total savings (action plan)	0.301 MMUSD
Electricity savings	0.29 MMUSD
Heating savings	0.012 MMUSD
Percentage of global savings	4.8%
Electricity percentage	12.1%
Heating percentage	2.7%

Barriers were like the other cases. However, in this case, they caused three companies to leave the network and four companies to fail to carry out the energy diagnosis proposed by the project team.

The last LEEN in the framework of the EEArg Project was the wine sector[9] network. This network was not initially planned by the project but given the scope and results of the five previous networks, the Energy Secretariat requested the European Union to finance an additional network.

Figure 4.16 Companies participating in Wine Sector LEEN.

Source: https://eficienciaenergetica.net.ar/industria_redes_de_aprendizaje.php

[9]For more information see: https://eficienciaenergetica.net.ar/img_publicaciones/09281641_SECTORVITIVINICOLA.pdf.

Initially, 22 companies joined the network, but due to the situation and many of the barriers faced, many companies abandoned the network without even receiving the energy diagnosis visit from the consultants. Only 11 companies completed the network (Figure 4.16). Like many previous ones, this network was developed entirely in a pandemic and isolated situation, so all eight workshops developed were virtual.

As can be seen in Table 4.5, the overall percentage of savings for this network was 2.5%, which for this network has been concentrated in the electrical source and, therefore, electrical savings. The source and its thermal savings have not been considered in the improvement potential because the warehouses practically do not use thermal fuels for their processes, or very little. For this network, energy costs represent an average of 15.7% of operating costs, which is very representative of the network's success.

Table 4.5 Overall results of the 11 companies that make up the Wine Sector Learning Network

Total energy costs	2,104 MMUSD
Electricity costs	2,046 MMUSD
Heating costs	0.058 MMUSD
% on operating costs	15.7%
Action plan (energy performance improvement)	18 months of participation in the Network
Investment	€0
Total savings (action plan)	0.052 MMUSD
Electricity savings	0.052 MMUSD
Heating savings	—
Percentage of global savings	2.5%
Electricity percentage	—
Heating percentage	2.5%

Source: https://eficienciaenergetica.net.ar/industria_redes_de_aprendizaje.php

4.4.3 Latest Networks in Argentina: National and Provincial Experiences Without Foreign Funding[10]

In November 2022, the Secretary for Energy launched the 8th LEEN in the City of Mar del Plata (Province of Buenos Aires). This network is part of an inter-administrative agreement between the Secretariat for Energy and the National Institute of Industrial Technology[11] (INTI for its acronym in Spanish). A key difference of this experience from other ones (both local and regional) is that it is financed with funds from the National Treasury. Thirteen companies belonging to different industrial sectors such as food, metalworking, fishing, and medical supplies, among others, participated in this initiative. All of them were trained in the implementation of an Energy Management System based on the international standard ISO 50001:2018 over the course of a year. According to official information on the website of the Argentinean Government[12], as a result of the training and exchanges in the workshops of this network, eight of these companies were able to implement a management system in their organizations, reaching 60% of its application jointly. These companies are La Campagnola, Nutreco, Palmar, QM Equipment, Laboratorios Quimicos, Compañía Industrial Frutihortícola, Tecmar, and Pesquera Veraz (Bonasur and Conservas Marechiare). Thanks to the energy efficiency measures detected in this process, these companies also achieved global savings close to 8% in terms of energy costs.

This 8th LEEN is the first experience developed and financed exclusively by the Secretary for Energy of Argentina, which has had the technical support of INTI for its realization, setting a good precedent in the consolidation of this work methodology in terms of public policies for the productive sector.

Then, in May 2023, in the framework of the project "*Mitigation of greenhouse gases and adaptation to the impacts of climate change in Latin America by strengthening energy efficiency in strategic sectors in Argentina and Chile*" financed by the EUROCLIMA+ Program, the Secretariat for Energy, the Ministry of Energy from Chile, and Chile's

[10]For more information (in Spanish) go to: https://www.argentina.gob.ar/economia/energia/eficiencia-energetica/eficiencia-energetica-en-sectores-productivos/redes-de-aprendizaje-de-eficiencia-energetica.

[11]https://www.argentina.gob.ar/inti

[12]https://www.argentina.gob.ar/noticias/finalizo-la-8va-red-de-aprendizaje-en-gestion-de-la-energia-de-mar-del-plata-y-alrededores.

Energy Sustainability Agency launched the Ninth Argentinean LEEN: The LEEN for Central Region, in the City of Cordoba[13]. According to information from the website of the EUROCLIMA+, this network corresponds to product 1 of this project, "I*ncorporating Energy Management Systems in 20 energy-intensive companies*". Product 1 is part of the line of action that seeks to develop instruments, mechanisms, and tools for the promotion and dissemination of energy efficiency. It seeks to generate demonstration effects in the private sector that can be replicated. Its objective is to increase energy efficiency in industrial and service companies through the incorporation of EMS in 20 energy-intensive companies in Argentina and Chile (10 in Argentina and 10 in Chile). Fifteen companies from the provinces of Córdoba, Santa Fe, Catamarca, La Pampa, Neuquén, and Buenos Aires, belonging to different industrial sectors such as food, metal-mechanics, aeronautics, and oil services, among others, are participating in this initiative.

The 10th energy management learning network run by the Secretariat of Energy, once again in conjunction with the National Institute of Industrial Technology (INTI), began on September 2023. In this case, it is a network for companies in the districts of San Martín and Tres de Febrero, in the province of Buenos Aires, and it is composed by 16 companies belonging to different productive sectors, including cosmetics, metalworking, textiles, and food. As stated by the Director of Renewable Energies from the Secretary for Energy "The aim of this network is to drive improvements in energy performance at low cost and without affecting the quality of their production, by learning about energy management concepts and tools completely free of charge."

In 2025, the Province of San Luis launched a new launched a new edition of the provincial learning network (a previous one was implemented in 2014). The initiative was promoted by the Secretariat of Environment of the Province of San Luis and INTI and includes technical advice and support in the implementation of energy efficiency measures. Under the slogan "Sustainability and Efficiency: savings for companies, benefits for the province,"

[13]https://www.agenciase.org/2023/05/23/se-lanzo-en-cordoba-red-de-aprendizaje-en-sistemas-de-gestion-de-la-energia-region-centro-en-el-marco-del-proyecto-de-euroclima-de-la-secretaria-de-energia-de-argentina-del-ministerio-de-e/

the initiative seeks to foster networking and knowledge exchange on energy efficiency, promoting resource savings and sustainable development. This network, which consists of 13 meetings, includes 30 companies.

4.4.4 On going LEENs in Small Enterprises: Patagonia Argentina: Energy Efficiency as a Transition Strategy for SMEs

In 2025, the project "Patagonia Argentina: Energy efficiency as a transition strategy for SMEs," implemented by the Argentine Confederation of Medium Enterprises (CAME)[14] and the Bariloche Foundation (FB)[15] and co-financed by the European Union was initiated. This project was built on the identification of problems and barriers to the implementation of energy efficiency in SMEs carried out by Fundación Bariloche and CAME during the Energy Efficiency Project in Argentina, within the framework of which work was done on the development of a National Energy Efficiency Plan Proposal (PlanEAr) that includes a specific chapter on analysis and recommendations for SMEs; and on the implementation of six RdA in SGEn in different regions/provinces of the country.

This proposal also incorporated lessons learned from the recent experience in Argentina mentioned in the previous sections. An important feature of this initiative is the close relationship between the implementing institutions and businesses, particularly CAME's track record and experience.

This project has at least two characteristics that diferentiate it from recent experiences in the country: it is exclusively for SMEs and is being implemented in a remote area of the country, the Patagonian region which is far from the main centers of activity.

The Argentine Patagonia is characterized by low population density and high territorial dispersion; while its surface area represents almost 50% of the country's total, only 6.4% of the population lives there. Another distinctive feature is the great distances between its cities, all of which have well under 500,000 inhabitants. At the same time, there are significant differences in the economic and socio-demographic base within the same province.

[14]https://www.redcame.org.ar/.

[15]https://fundacionbariloche.org.ar/.

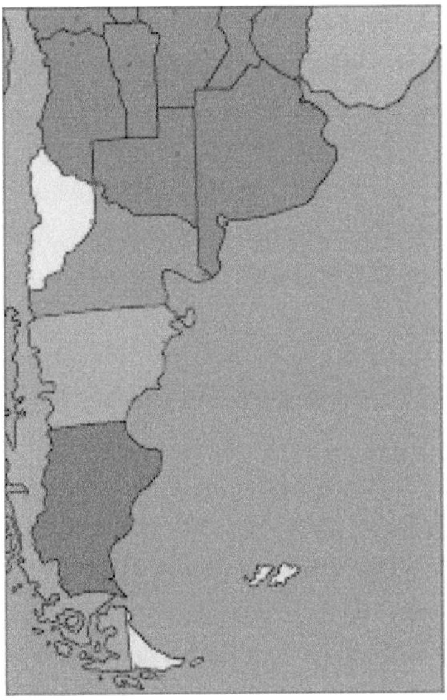

Figure 4.17 Provinces belonging to the Patagonia Argentina.

In Argentina, there are 600,000 small and medium-sized enterprises distributed throughout the country. Despite their socioeconomic relevance SMEs face serious sustainability problems. For instance, they deal with difficulties in accessing financing, which mainly translates into a lack of investment and infrastructure Argentine SMEs account for 33% of the country's industrial sector energy consumption, including self-production (48% of electricity, 34% of GD, and 8% of other sources) (Bouille et al., 2019). This is crucial when analyzing energy efficiency strategies. In addition, studies conducted as part of the "Energy Efficiency in Argentina" project indicate that, while barriers to energy efficiency are similar to those faced by large companies, the role of barriers related to technological and capacity capabilities and information barriers is more relevant (Bouille et al., 2021). In the case of technological and capacity barriers, the problem relates to the internal capacity to evaluate options and alternatives for energy efficiency actions, as well as the evaluation of financing possibilities. In the case of financing barriers, it relates to the problem of access to financing and its cost.

In this context, the implementation of energy efficiency measures in SMEs is a great opportunity to improve the economic condition of companies, while generating the co-benefit of contributing to the sustainability of the energy system and the mitigation and adaptation of the sector. However, these measures are not frequently implemented in Argentinean SMEs.

For this reason, the project aims to promote the creation of four LEENs in the different Patagonia provinces, each one composed of 10-12 SMEs from different cities and economic activities of the provinces.

Building local capacities in a remote region

A distinctive and fundamental aspect of this project is that it included a training component for professionals in energy management in the Patagonian region. The project proposes the delivery of virtual courses that ended with a face-to-face session and the awarding of certificates. The course consisted of 30 classes, 120 hours of training. It was taught by an international expert in Energy Management and Learning Networks.

According to the project implementers, this action will have a direct and positive impact on professionals and will indirectly contribute to the growth of green jobs in the region beyond the direct impact on these LEENs. The opinions of the professionals that participated in the course show that it was a very important strategy to increase regional capacities.[16]

In the first of the course between March 2025 and June 2025, 55 Patagonian professionals finished successfully the formation from which 32% were female.[17]

The first LEEN of the Patagonia Argentina Project: Rio Negro

The first of the fourth LEENs to be developed in the context of the "Patagonia Argentina: Energy Efficiency as a Transition Strategy for SMEs Project" was in the province of Rio Negro. It was initiated in June 2025 and ended with the Fourth training workshop in October 2025. It was composed by 11 SMEs belonging to diverse industrial sectors and different cities and towns of the province: Agro Fresh S.A.

---

[16]Please visit: https://www.instagram.com/reel/DL0JYTFAMN9/?igsh=MWJpaGFreXRmY3NscA%3D%3D
https://www.instagram.com/reel/DF5zH1WxO-q/?igsh=Zmd0Z3V2cnQ1bXZm

[17]https://www.redcame.org.ar/novedades/14265/generacion-de-empleo-verde-en-la-patagonia-argentina-came-y-la-fundacion-bariloche-formaron-80-nuevos-auditores-energeticos

and LP SRL (General Roca); Bonadé S.A., Polano S.A. Frutas Stelzer and Jugos S.A.(Villa Regina); Deco Klart (General Fernández Oro); Hacienda Martínez (Ingeniero Luis A. Huergo); Rimini Hotel and Supermercado Palito (Las Grutas) and y Villa Huapi (Dina Huapi).

As stated by the implementers, in this first edition of the LEENs for SMEs of the region, the main challenges were limited time and resources for data acquisition and action plan development, the need to rethink the business model, changes in management behavior, and the lack of an energy performance culture, among others.

The participants highlighted that the experience was very relevant for them and that, thanks to the LEEN, they could incorporate new management tools, improve the organization of data and priorities, and identify opportunities to prioritize energy improvements, among others.[18]

In November, the second LEEN was launched, this time in the province of Neuquén. As it is still being set up, there is no information yet on the companies that will form part of it.

4.5 Main Results of Argentinean LEENs

Most of the quantitative information on Argentinean LEEN results analyzed in this section comes from the EE Argentina.

The results seem to have been very satisfactory, both for the industries that have participated in the programs and for professionals who have attended all the learning workshops. For the industries, it has been a starting point to give a standardized approach to energy management in their organizations. Professionals have worked on different energy analysis tools that will allow them to grow in their respective jobs, achieving added motivation to their usual responsibilities.

The global results of two of the six completed EU-financed LEENs are shown in Table 4.6. The total energy cost of the companies involved in these LEENs were 229.91 MMUSD, which represents an average operating cost of 11.7%; while the total potential savings estimated were 5.54 MMUDS, which represents a percentage of nearly 2.5%. The results of each of the LEENs of the Argentina Energy Efficiency Project are also shown in Table 4.7.

[18]For more information, please visit: https://shorturl.at/F1oFu; https://shorturl.at/eAl0B.

Table 4.6 Overall results of the six EU LEENs

Total energy costs	**228.9 MMUSD**
% on operating costs	11.7%
Action plan (energy performance improvement)	18 month of participation in the Network
Investment	0
Total savings (action plan)	5.54 MMUSD
Percentage of global savings	2.42 %

Source: Based on information from the Energy Efficiency in Argentina Project.

Beyond the quantitative results of the LEENs, other qualitative impacts are very important in terms of cultural changes and information generated. Indeed, probably one of the greatest contributions from LEENs is the cultural change among the professionals of the firms, which included recognition of the importance of the measurement of energy consumption. This has been mentioned in different situations by the participants. Benefits also include the interaction of the industries and the exchange of knowledge. The following paragraphs resume the main opinions of some of the participants of the EU-financed LEENs and the evaluation elaborated by the coordination (technical experts and the professionals from the Secretary for Energy) of the LEENs[19].

The LEENs allowed companies to implement energy management practices, improving its operational controls related to significant energy uses; improving and deepening the previous analyses and understanding on topics such as energy review, energy indicators, and baselines; maximizing the use of existing elements in the organization, whether from other management systems such as operational and maintenance; demonstrating improved energy performance; measuring energy savings; and participants internalized in the use and application of simple tools that allowed them to generate new changes. Synergies were also generated beyond energy efficiency and management in the sense that during the meetings.

[19]These comments have been mainly gathered from the two networks that had an in-person closure (Santa Fe and Tucumán); the other LEENs could not be closed in person due to the restrictions imposed by the COVID-19 pandemic.

Table 4.7 Results of each of the six EU-financed LEENs

	Santa Fe	**Tucumán**	**Córdoba**	**Pilar**	**Misiones**	**Wine**
No. of companies	10	13	11	11	11	11
Total energy cost	131.4 MMUSD	57.7 MMUSD	14.6 MMUSD	20.37 MMUSD	2.834 MMUSD	2.104 MMUSD
Electricity costs	80.4 MMUSD	16.6 MMUSD	12.7 MMUSD	19.61 MMUSD	2.396 MMUSD	2.046 MMUSD
Heating costs	50.4 MMUSD	41.1 MMUSD	1.9 MMUSD	0.76 MMUSD	0.436MMUSD	0.058 MMUSD
% on operating costs	20.8%	21.8%	6.9%	11.5%	13.5%	15.7%
Action plan (energy performance improvement)	18	18	18	18	18	
Total savings (action plan)	2.7 MMUSD	1.5 MMUSD	0.65 MMUSD	0.339 MMUSD	0.301 MMUSD	0.052 MMUSD
Electricity savings	1.8 MMUSD	0.3 MMUSD		0.323 MMUSD	0.29 MMUSD	0.052 MMUSD
Heating savings	0.9 MMUSD	1.3 MMUSD	0.4 MMUSD	0.011 MMUSD	0.012 MMUSD	
% of global savings	2	3	3.6	1.6	4.8	2.5
% of electricity	2.3	1.7	4.8	1.7	12.1	
% heating	1.7	3.1	2.1	1.5	2.7	2.5

Source: Based on information from the Energy Efficiency in Argentina Project.[20]

[20]Values estimated for an exchange rate of Argentine peso / Dollar 60.5 AR$/USD—average buyer and seller. *Source*: Banco Nacion, November 2019.

For professionals, the LEENs have also been an interesting experience, which seems to have enhanced their professional capabilities. As part of the evaluation process, of the LEENs, participants were asked to provide their opinions and experiences.

4.6 Lessons Learned and Challenges from Argentinean Experience

Industrial energy consumption is very relevant for the Argentinean energy sector, and it is also very important in terms of environmental impact. Nonetheless, different enabling conditions (low-energy prices, lack of energy efficiency, well-targeted policies, among others) have led industrial energy efficiency as a non-relevant topic for companies, up to the moment.

In this context, the LEENs have emerged, firstly because of a triangular cooperation, with a significant role of Mexico and GIZ, then financed by the European Union and EUROCLIMA+ program, and more recently entirely financed by the local government to incentivize the implementation of energy efficiency actions in this sector. These LEENs have been found as an effective alternative to remove some of the main barriers confronted by the industrial sector: the information and capacities barriers (as well as financing problems).

Regarding the local configuration of the networks, it is important to highlight some of the main characteristics of all the LEEN implemented so far: (a) include a mixture of large companies and SMEs and (b) include companies of several industrial sectors.

One of the most important barriers faced in the implementation of LEENs was the lack of energy information. As mentioned by the energy consultants, due to the energy culture of organizations energy consumption is broken down by different departments with low direct connection. This has been indeed an important barrier to the implementation of the LEENs.

In addition, in those industries where CEOs have been aligned with the objectives of the networks from the beginning, better results have been achieved in energy savings and learning outcomes for professionals. This aspect is essential, and the networks implementation team has designed a strategy for meetings with the different CEOs of the different networks, where the aim is to show the advanced work and the benefits offered by the implemented format.

Within the successful aspects of the LEEN experiences in Argentina, the implementation of the workshops must be highlighted. This includes the design of visualization techniques and exchange dynamics. Workshops provided an excellent space for the exchange of knowledge that strengthened the capacities of each of the members, generating trust, and cohesion, and establishing a relationship. At the end of each workshop, a space was opened to participants to identify aspects to improve in the workshop, to adjust both logistics, on the part of the technical team and participants on how to contribute and create a better space for interaction to take advantage of the exchange of knowledge in an innovative and enriching way. In addition, communication mechanisms were fast, easily accessible, and able to maintain the link between the group in a transparent and expeditious manner in relation to queries related to the EMS.

As previously mentioned, the literature shows that a good performance of LEENs requires some favorable boundary conditions. In the Argentinean case, the main boundary conditions were: (a) the energy consumption for the company participating in the firm budget (according to the company criteria); and (b) the commitment of the management of the company to implement an EMS (Carpio and Recalde, 2020).

As argued by Carpio and Recalde (2020), these are some lessons learned from the Argentinean experience that should be considered in future (local or regional) extensions of these networks from the very beginning of the project:

- Engage actors with decision-making power in each company to raise awareness.
- Have a shared repository of information generated and/or used within the network.
- Encourage the development of some elements of the EMS in a collaborative way, with contributions and experiences of the same participants.
- Undertake an investment plan in measuring equipment from the beginning, to be able to quantify energy savings

opportunities; care needs to be taken to ensure that these investments are frugal and appropriate for the most important measurements, to avoid unnecessary over-measurement.

- Emphasize the training process of the technical teams in each industry.
- Make technical information available through a variety of approaches including new information and communication technologies.
- Strengthen the exchange between the different learning networks not only intra-network since the objectives and learning are shared.
- It might be beneficial to create a space to share progress and lessons learned, generated in the same process of the network.

Finally, a novel and interesting aspect of this experience is that its scope has been scaled up to the point that the energy ministry has implemented a large number of networks in conjunction with INTI, while other initiatives have been implemented by municipal governments and SMEs have recently been included as a target population. This aspect highlights the importance that this mechanism has gained at the national level.

Chapter 5

Energy Efficiency Learning Networks in the Brazilian Water Sector

Rita Cavaleiro de Ferreira
Strong Water Utilities in Jordan, GITEC-IGIP GmbH, Germany
rita.cavaleiro@gic-group.com

5.1 Introduction

This article summarizes the experience of learning networks that occurred between 2017 and 2021 in the water and sanitation sector in Brazil through the "**Energy Efficiency in Urban Water Supply Systems Project (ProEESA)**"[1]

Seven learning networks were carried out within the German–Brazilian cooperation having as partners on the Brazilian side—the National Sanitation Secretariat of the Ministry of Regional Development (SNS/MDR), and on the German side—the Federal

[1]In Portuguese—Projeto de Eficiência Energética no Abastecimento de Água-ProEESA.

Learning Energy Efficiency Networks: Applications and Challenges in Latin American Countries
Edited by Claudio Carpio and Marina Yesica Recalde
Copyright © 2026 Jenny Stanford Publishing Pte. Ltd.
ISBN 978-981-4968-82-9 (Hardcover), 978-1-003-77468-6 (eBook)
www.jennystanford.com

DOI: 10.1201/9781003774686-6

Ministry for Economic Cooperation and Development (BMZ), through the GIZ.

This chapter covers(i) the context of efficiency in public utilities in Brazil, (ii) the concept of applied learning networks to identify major differences and variations regarding learning networks that occurred in other countries, (iii) the description of seven learning networks and their main achievements, (iv) some approaches used to increase participants commitment in the specific sector setting of public utilities, and (iv) publications produced associated to energy efficiency in the learning networks of the water and sanitation sector.

5.2 The Context of Efficiency in Public Water and Sanitation Utilities

5.2.1 Barriers to Energy Efficiency

Barriers to energy efficiency have multiple aspects such as low energy prices, counterproductive subsidies given to special sectors, lack of know-how in efficiency management, user aversion to more efficient technology, and barriers to huge initial investments and regulatory impediments.

In the first six of the executed learning networks, ProEESA addressed mainly the knowledge and management barriers to efficiency. Later, an important energy subsidy to water service providers was gradually reduced over a 5-year period (starting in 2019). Despite much of the sector's reluctance, this decision of the Brazilian Ministry of Energy and Mining increased the interest in energy efficiency of water service providers.

In the seventh learning network, ProEESA addressed water and energy efficiency in a more structural way through regulation and by approaching a more institutional setting for efficiency and friendly operations and investments. The outputs of this last learning network seemed to be more abstract but surprisingly several regulation agencies incorporated some of the discussed and tested tools and started to induce immediately more efficient decisions and behavior on the water utilities side.

5.2.2 Characteristics of Public Water Supply and Sanitation Sector

Public water supply and sanitation services are characterized by being a natural monopoly. This means that usually there are not two or more water providers competing and offering their services in one urban area. For industries that rely heavily on physical networks (water supply and wastewater collectors), it is not affordable to have more than one network in a city. In this way, the public water and sanitation service is a natural monopoly.

This is a significant difference from other industries that compete to offer the lowest price and have a real incentive to pursue cost reductions and ways to expand the efficiency frontier.

For the water and sanitation sector, this attribute of natural monopoly means that there are less favorable conditions for the continuous search for efficiency expansion. In this context, there is no real need to explore the frontiers of efficiency because service providers can cover budget deficits with tariff rebalancing, either by increasing water tariffs or relying on public taxpayer-funded subsidies.

Some experts may defend that competition for efficiency (inclusive of energy efficiency) occurs when a tender for water service provision is launched. In this situation, several private bidders offer service provision with a predetermined tariff, and the service is awarded to the bidder with the lowest price. This procedure is often seen to some extent as the "privatization of public services" for improved efficiency. Nevertheless, when these bids occur, they tend to have long concession periods of more than 15, 20, 30, or 40 years, in which operational and infrastructural efficiency tend to be neglected during these long periods.

So, in natural monopolies, efficiency negligence occurs regardless of the legal nature of the operator, whether it is a public, private, or mixed entity. Therefore, regulation must step in and try to identify inefficiencies that are not transposed to the tariff and taxpayer. Regulation plays a key role in setting efficiency incentives, and it occurs with different levels of information and decision-making transparency.

5.2.3 The Brazilian Institutional Setting of the Water Sector

The Brazilian setting of the water sector is extremely heterogeneous. There are 5668 municipalities responsible for water and sanitation, providing the service directly or through delegation to external private or public companies. 1341 service providers responded to the annual water and sanitation survey (SNIS[2]). The following graphic shows the amount and size of service providers. 63% of the Brazilian population lives in the service area of one of the 24 very large water and sanitation companies. The biggest one is SABESP in the estate of Sao Paulo with almost 30 million inhabitants (about 14% of Brazil's population). Figure 5.1 shows the frequency of the following sizes of companies:

- Very small water utilities—serving up to 15,000 inhabitants
- Small water utilities—serving between 15,000 and 50,000 inhabitants
- Medium size water utilities—serving between 50,000 and 200,000 inhabitants

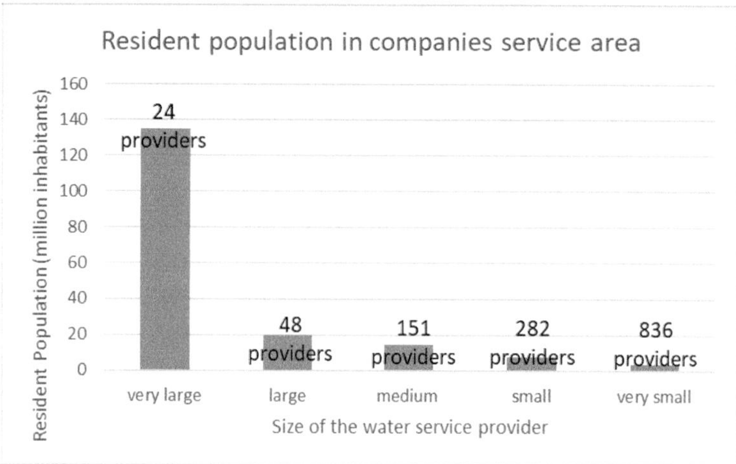

Figure 5.1 Size and frequency of water and sanitation utilities in Brazil.
Source: SNIS 2020 (http://app4.mdr.gov.br/serieHistorica/).

[2]http://www.snis.gov.br/.

- Large water utilities—serving between 200,000 and 1,000,0000 inhabitants
- Large water utilities—serving over 1,000,0000 inhabitants

The heterogeneity of the size of the companies has reflections on the qualifications and skills of their human resources. The main scope of the learning networks was large, medium, and small service providers. Nevertheless, in one case of a very large service provider (Sanesul with 1.5 million inhabitants), a learning network was conducted within the company with its several regional units.

5.3 Learning Network Concept

The Learning Network approach is applicable to any topic such as Municipal Basic Sanitation Plans, Water Safety Plans, energy efficiency, water losses, asset management, tariff and cost calculation, sewage treatment, urban rainwater drainage, and information management systems, just to name a few, of relevance within the water sector. The topic must be of common interest amongst a group of organizations. There are no limits to the themes to be addressed in the Learning Networks; however, a clear delimitation of scope is required to maintain the focus and the reason for the learning network to exist in the first place.

In Brazil, for the water sector, the Learning Network approach was set up to address specifically energy efficiency. Its format was a medium-term training program (12–18 months) based on face-to-face meetings with works and tasks to be carried out throughout this period and an expected output product. This specific training and experience-sharing format brings into the group external knowledge input and shares best practices of participants that encourage the other participants too.

This approach is considered to have high permeability within the adherent organizations, an improved pedagogical environment, enhanced networking across participants, monitoring, and accountability by specialists. These characteristics improve the chances of real and enduring change that will impact the efficiency levels of the utilities in the short, medium, and long term.

In a Learning Network, there are three key roles:

- The Anchor entity and coordinator of the initiative organizes the program and the content. Usually, the regulation agency or the headquarters of a large water utility assumed this role.
- Participating service providers (utilities or regional units in case of very large companies) with their technicians.
- External specialists in key topics relevant to learning network output.

In the Brazilian version of the learning networks for the public sector, no admission fees were collected. Costs covered coordination and services of external consultants (when applicable) were provided by the German Brazilian Cooperation and the local anchor entity. Beneficiary organizations provided venue facilities hosting the meetings and provided input time for their staff.

5.4 Learning Networks Description

Seven learning networks were held in five years during 2017 and 2021 in the Brazilian water and sanitation sector (Figure 5.2). All networks had as overall institutional partners on the Brazilian side—the National Sanitation Secretariat of the Ministry of Regional Development (SNS/MDR) and on the German side—the Federal Ministry for Economic Cooperation and Development (BMZ), which supported the networks with resources through the GIZ.On the operational level, regional partners were involved besides the beneficiary organizations.

The first five networks had similar scopes, focusing on the elaboration of **Water and Energy Management Plans** on a municipal level with a 20-year time horizon including a 3- and 10-year action horizon. These networks have similar formats, such as regular classroom meetings gathering up to 15 organizations with up to three participants each, in a nearby geographic area, mainly coordinated by a regional entity and the local regulation agency (or the state service provider).

Only a few of the identified energy efficiency measures were possible to implement during the learning networks period, especially the more operational ones that did not require major investments. These low-hanging fruits originated immediately and have considerable cost and energy savings in some companies. All other efficiency measures that required a significant amount of investment needed further studies and financial resources besides extended tender procedures to execute the measures. In this sense, the radical efficiency impact was not measurable during the short period of the learning networks.

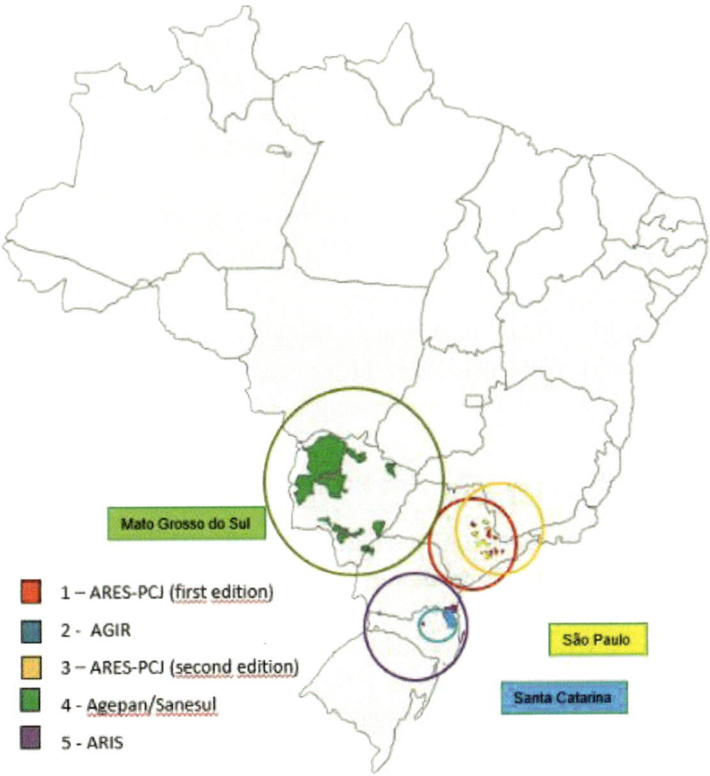

Figure 5.2 Location of five face-to-face learning networks (2018–2020).

Source: Own elaboration.

The sixth and seventh learning networks differed significantly in scope, format, and size. Both were held remotely online and covered over 50 participant organizations across South America during the pandemic restriction period. GIZ coordinated with appropriate nationwide sector entities, the network content, and the meetings.

The expected outputs of the sixth learning network focused on following up and monitoring measures identified by water operators that participated in earlier editions while incorporating newcomers too.

The outputs of the seventh learning network focused on tools to regulate efficiency, from the regulation agencies perspective, while addressing also, water operators and sector consultants. Water and energy efficiency was addressed in a more structural way, through regulation and approaching the wider institutional settings. The outputs of this last learning network seemed to be more abstract. Surprisingly several regulation agencies managed to pass regulations with some of the discussed and tested tools inducing immediate effects on the water utilities mentioned in the following descriptions of the seven networks.

Below are brief descriptions of the seven learning networks mentioned:

- Main goal and execution period
- Institutional partners and participants
- Special characteristics of the learning network
- Main outputs and some relevant side effects

5.4.1 ARES-PCJ, 2017, São Paulo Estate

Main goal and execution period: The main goal of this learning network was to structure **Water and Energy Management Plans**, on the municipal level including the election of KPI, baseline, goals, and calculation of saving potential and identification of efficiency projects. The **ProEESA Learning Network on Water and Energy Management Plans** was held between February 2017 and January 2018.

Operational partners and participants: The local partner was the Regulation Agency for the Water and Sanitation Services in the basins of the Rivers Piracicaba, Capivari e Jundiaí in São Paulo Estate (*ARES-PCJ –Agência Reguladora dos Serviços de Saneamento das Bacias dos Rios Piracicaba, Capivari e Jundiaí*). The beneficiaries were the water and sanitation service providers of 12 municipalities[3]with 41

[3]Araraquara; Araras; Atibaia; Cerquilho; Indaiatuba; Leme; Louveira; Mogi Mirim; Piracicaba; Santa Bárbara de Oeste; Valinhos; Vinhedo.

technicians. These participating municipalities serve a population of 750,000 inhabitants.

Special characteristics of this learning network: The preparation of the learning network was significantly longer than the following ones (about one year) because it was the first that all pedagogic handouts had to be prepared and adjusted to the Brazilian institutional framework.

The participants of this learning network were selected through a call with the regulation agency, where the most committed was prioritized.

For this learning network, some financial resources even made the acquisition of measurement equipment possible (flow meter, energy quality meter, and pressure meter), which was used to survey the efficiency levels of several pumping stations and calculate the accurate saving potential, necessary to ensure the return and viability of the investment.

After the learning network, the measurement equipment was handed out to the regulation agency ARES-PCJ, where the regulated municipalities used and borrowed it like books from a library.

For up scaling purposes of the learning network approach, another regulation agency (AGIR in Santa Catarina Estate) was invited to witness the process and evaluate whether such an initiative would be interesting in their intervention area. AGIR found it relevant, and later in the same year, a learning network was initiated.

Main outputs:

- On the documentation level by the participants, the main outputs were:
    - Three concluded water and energy loss control plans with sufficient quality level
    - Five partial plans with the main topics covered
    - Four exercises/preparatory studies and reflections on water and energy loss
- Overall saving potential identified was 16% of operational costs (91,965,000 R$/year) amongst participants:
    - Saving potential identified in the field of fraud control and water meter under measurement (54,645,000 R$/year)

- Saving potential identified in the field of real water losses—avoidable production costs (35,338,000 R$/year)
- Saving potential identified in the field of electromechanical pumping efficiency—avoidable electricity costs (4,792,000 R$/year)
- Saving potential identified in the field of avoiding penalties and optimizing power factor—avoidable electricity costs (1,583,000 R$/year)
- Saving potential identified by transferring operating hours to off-peak hours—avoidable electricity costs(399,000 R$/year)

- Overall natural resources saving potential identified:
  - Avoidable real water losses 7% of produced water (16,326,000 m^3/year)
  - Avoidable electricity consumption 14% (27,310,000 kWh/year)
    - Through water loss reduction (13,680,000 kWh/year)
    - Through improved electromechanically efficiency (13,631,000 kWh/year)
  - Avoidable CO_2 emission 2,532 tons/year[4]
- Internal organization adaptations to address explicitly water and energy losses in three water and sanitation service providers[5] include establishing specific commissions and defining new competencies.
- In one Municipality, Atibaia, technicians were fit to prepare a financially viable energy efficiency project according to the PMIVP and the requisites of PEE.

A short video was made by ARES-PCJ to report on this learning network, with English subtitles (Figure 5.3). https://www.youtube.com/watch?v=5wUMYOtSk4E&t=90s with Portuguese subtitles.

---

[4]This figure is very low compared to other countries because Brazil's energy matrix is one of the cleanest globally.

[5]Commission for Water and Energy Loss Reduction in Araras Municipality; Commission for Water and Energy Losses—Municipality of Valinhos); Expansion of functions in the Command and Control Center in the Energy Monitoring System—implemented in the municipality of Indaiatuba.

Figure 5.3 short video was made by ARES-PCJ to report on this learning network.

Source: https://www.youtube.com/watch?v=C_Ko2zJsd5U.

5.4.2 AGIR, 2017, Santa Catarina Estate

Main goal and execution period: The main goal of this learning network was to structure **Water and Energy Management Plans**, on the municipal level, including the election of KPI, baseline, goals, and calculation of saving potential and identification of efficiency projects. The **ProEESA Learning Network on Water and Energy Management Plans** was held between November 2017 and October 2018.

Operational partners and participants: The local partner was the Regulation Agency for the Water and Sanitation Services in the middle basins of the River Itajaí in Santa Catarina Estate (*AGIR—Agência Intermunicipal de Regulação do Médio Vale do Itajaí*). The beneficiaries were four water and sanitation service providers operating in 15 municipalities[6] with 31 technicians participating. These participating municipalities serve a population of 1,000,000 inhabitants.

---

[6]Benedito Novo, Apiúna, Blumenau, Ascurra, Gaspar, Botuverá, Brusque, Guabiruba, Pomerode, DoutorPedrinho, Indaial, Rio dos Cedros, Rodeio, Timbó, Guaramirim.

Special characteristics of this learning network: The participants of this learning network were invited. The agency regulates 14 municipalities considered to be a good group size of participants. Neither call nor tender was made to participate. Additionally, a neighbor regulation agency (ARIS in Santa Catarina Estate) and one of their regulated service providers (Guaramirim) were invited to witness the process to evaluate if such a learning network would be interesting to be conducted in their intervention area. It turned out that ARIS was interesting and later in the next year, an own learning network was initiated.

Main outputs:

- On the documentation level by the participants, the main outputs were:
  - Eight concluded water and energy loss control plans with sufficient quality level
  - Six partial plans with the main topics covered
  - One preparatory study and reflections on water and energy loss were made
  - 214 water and energy efficiency projects identified with an investment of 280,000,000 R$ to be implemented over 10 years (31 regarding real water losses, 26 regarding apparent water loss, 60 regarding electro-mechanical efficiency, 25 regarding improved asset management, and 1 regarding water saving measures)
- Overall saving potential identified was 20% of operational costs (33,167,000 R$/year):
  - Saving potential identified in the field of fraud control and water meter under measurement (13,372,000 R$/year)
  - Saving potential identified in the field of real water losses—avoidable production costs (16,436,000 R$/year)
  - Saving potential identified in the field of electro-mechanical pumping efficiency—avoidable electricity costs (2,272,000 R$/year)
  - Saving potential identified in the field of avoiding penalties and optimizing power factor—avoidable electricity costs (101,000 R$/year)

- Saving potential identified by transferring operating hours to off-peak hours—avoidable electricity costs (986,000 R$/year)
- Overall natural resources saving potential identified:
  - Avoidable real water losses 7% of produced water (4,706,000 m^3/year)
  - Avoidable electricity consumption 12% (34,899,000 kWh/year)
    - Through water loss reduction (13,680,000 kWh/year)
    - Through improved electromechanically efficiency (4,346,000 kWh/year)
  - Avoidable CO_2 emission 2,358 tons/year[7] (this figure is very low compared to other countries because Brazil's energy matrix is one of the cleanest globally)

Important side effects:

- The regulation agency AGIR is a very dynamic agency and offering initiatives like the learning network makes it more valuable to its associated/regulated service providers.
- Some quick changes were made to the operational and analytical level that improved immediately the efficiency. One service provider[8]upgraded the analysis procedures of its data calculatingonlinepumpingefficiency.Inasecondmunicipality,[9] network pressure was lowered overnight to reduce real water losses. Two other municipalities[10]implemented water meter district zones for improved water loss monitoring and control.
- In one Municipality, Blumenau, technicians were fit to prepare the necessary documentation for a financially viable energy efficiency project according to the International Performance Measurement and Verification Protocol (MVP) and the requisites of the Energy Efficiency Programme (PEE) participating successfully in the Energy Service Provider (CELESC) Program Bonus Motor, and in one year halving remarkably their energy costs and consumptions.

---

[7]This figure is very low compared to other countries because Brazil's energy matrix is one of the cleanest globally.

[8]SAMAE Brusque.

[9]CASAN Botuverá.

[10]CASAN Ascurraand CASAN Rio dos Cedros.

A small video was made to report this learning network (https://www.youtube.com/watch?v=s9iEktdh1R8) (Figure 5.4).

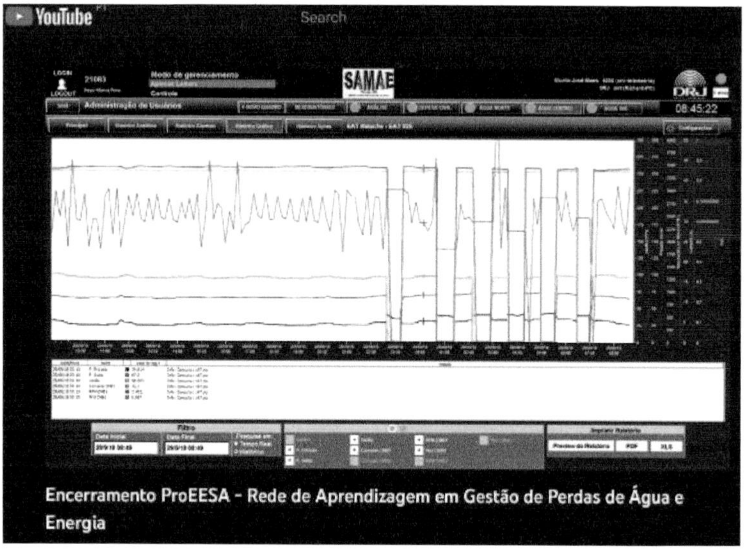

Figure 5.4 Video reporting the AGIR learning network.

5.4.3 ARES-PCJ, 2018, São Paulo Estate

Main goal and execution period: The main goal of this learning network was to structure **Water and Energy Management Plans**, on the municipal level, including the election of KPI, baseline, goals, and calculation of saving potential and identification of efficiency projects. The ProEESA Learning Network on Water and Energy Management Plans was held between June 2018 and April 2019.

Operational partners and participants: The local partner was the Regulation Agency for the Water and Sanitation Services in the basins of the Rivers Piracicaba, Capivari e Jundiaí in São Paulo Estate (*ARES-PCJ –Agência Reguladora dos Serviços de Saneamento das Bacias dos Rios Piracicaba, Capivari e Jundiaí—Estado de São Paulo*). The beneficiaries were the water and sanitation service providers of 10 municipalities[11] with 30 technicians. These participating municipalities serve a population of 2,000,000 inhabitants.

[11]Americana, Jundiaí, Pirassununga, Sorocaba, Brotas, São Carlos, Nova Odessa, Santa Rita do PassaQuatro, Porto Feliz, Itu.

Special characteristics of this learning network: This learning network was the second edition by this regulation agency, and therefore, was conducted by ARES-PCJ staff with internal know-how access to some external consultants. GIZ staff were in a coaching position with minor input. The output report was made and kept internal to ARES-PCJ.

A small video was made by ARES-PCJ to report on this learning network, with Portuguese subtitles (https://www.youtube.com/watch?v=0uC1WRhxT44) (Figure 5.5).

Figure 5.5 Video reporting the second edition of the ARES-PCJ learning network.

5.4.4 AGEPAN-Sanesul, 2018, Mato Grosso do Sul Estate

Main goal and execution period: The main goal of this learning network was to structure **Water and Energy Management Plans,** on the municipal level, including the election of KPI, baseline, goals, and calculation of saving potential and identification of efficiency projects. The ProEESA Learning Network on Water and Energy Management Plans was held between July 2018 and May 2019.

Operational partners and participants: The local partner was the Estate Regulation Agency for the Water and Sanitation Services in Mato Grosso do Sul (*AGEPAN –Agência Estadual de*

Regulação de Serviços Públicos de Mato Grosso do Sul) and the Estate Water and Sanitation service provider of Mato Grosso do Sul (*SANESUL—Empresa de Saneamento Básico de Mato Grosso do Sul*). The beneficiaries were 10 regional operation units of Sanesul serving 15 municipalities[12] with 44 technicians. These participating municipalities serve a population of 1,000,000 inhabitants.

Special characteristics of this learning network: All participants were from the same company, which operates in a very vast territory, almost the whole estate of Mato Grosso do Sul. To optimize traveling, the meetings needed an overnight stay and were 1.5 days long. The participating regional units were chosen by the SANESUL headquarters according to the most promising teams.

Cooperation among Regional Units was high. Good approaches from one regional unit were easily transferred to others. Also, Sanesul Headquarters made important contributions by establishing standard corporation standards and templates.

Main outputs:

- On the documentation level by the participants, the main outputs were:
    - 15 (out of 15) water and energy loss control plans with sufficient quality levels were concluded. This was the best output ever considering all previous learning networks.
- Overall saving potential identified was 35% of operational costs (87,700,000 R$/year):
    - Saving potential identified in the field of fraud control and water meter under measurement (21,000,000 R$/year)
    - Saving potential identified in the field of real water losses—avoidable production costs (49,200,000 R$/year)
    - Saving potential identified in the field of electromechanical pumping efficiency—avoidable electricity costs (6,400,000 R$/year)
    - Saving potential identified in the field of avoiding penalties and optimizing power factor—avoidable electricity costs (138,800 R$/year)

[12]Dourados; Maracajú; Rio Brilhante; Três Lagoas; Coxim; Jardim Sidrolândia; Naviraí; Nova Andradina; Amambai; Ponta Porã; Chapadão do Sul; Corumbá; Ladário; Aquidauana.

  - Saving potential identified by transferring operating hours to off-peak hours—avoidable electricity costs (5,900,000 R$/year)
- Overall natural resources saving potential identified were:
  - Avoidable real water losses of 14% of produced water (9,255,000 m^3/year)
  - Avoidable electricity consumption 58% (20,300,000 kWh/year)
    - Through water loss reduction (3,500,000 kWh/year)
    - Through improved electromechanically efficiency (16,700,000 kWh/year)
  - Avoidable CO_2 emission 1.887 tons/year[13]

Important side effects: The regular meetings during the learning network improved the company's cohesion and improved communication across operation units in such a vast territory. Headquarters improved communication with its regional units. Important approaches regarding the water balance and energy bill management have improved and converged.

5.4.5 ARIS, 2018, Santa Catarina Estate

Main goal and execution period: The main goal of this learning network was to structure **Water and Energy Management Plans,** on the municipal level, including the election of KPI, baseline, goals, and calculation of saving potential and identification of efficiency projects. The **ProEESA Learning Network on Water and Energy Management Plans** was held between November 2017 and October 2018.

Operational partners and participants: The local partner was the Regulation Agency for the Water and Sanitation Services in the State of Santa Catarina (*ARIS -Agência Reguladora Intermunicipal de Saneamento—Estado de Santa Catarina*). The beneficiaries were the water and sanitation service providers of 15 municipalities[14] with 62 technicians. These participating municipalities serve a population of 749,000 inhabitants.

---

[13]This figure is very low compared to other countries because Brazil's energy matrix is one of the cleanest globally.

[14]Itapiranga; Schroeder; Rio Negrinho; Tijucas; Três Barras; Joinville; São José do Cedro; Fraiburgo; Governador Celso Ramos; São Bento do Sul; São João Batista; Jaraguá do Sul; Meleiro; Palhoça; Florianópolis.

Special characteristics of this learning network: The participants of this learning network were invited by the regulation agency. There was a major input of the regulation agency in organizing the network content and programme since the German–Brazilian Cooperation ceased its activities in the middle of the learning network (May 2019). The regulation agency witnessed the learning network in its neighbor (AGIR) and conducted all activities.

Main outputs:

- On the documentation level by the participants, the main outputs were:
  - Five concluded water and energy loss control plans with sufficient quality level
  - Nine partial plans with the main topics covered

5.4.6 ProEESA 2.0, 2020, Nationwide, Brazil

Main goal and execution period: The main goal of this learning network was to support participants of previous learning network editions (nationwide) to (i) improve or extend the scope of their **Water and Energy Management Plans**, on the municipal level, (ii) elaborate monitoring reports based on the plans previously done, (iii) prepare studies of energy efficiency for the new municipal management 2021–2024, and (iv) adopt best practices and procedures within the organization related to water and energy information management according to the ACERTAR auditing methodology. The **ProEESA 2.0 Learning Network on Water and Energy Losses** was held between March and December 2020.

Operational partners and participants: The local partner was the National Sanitation Secretariat of the Ministry of Regional Development (SNS/MDR). The beneficiaries were the water and sanitation service providers of 65 service providers[15] with 218 technicians from 1146 municipalities.

[15]BRK Ambiental Limeira S/A; BRK Ambiental Sumaré S/A; CAEPA—Companhia de Água e Esgotos de Paraíbuna, São Paulo State; CAESB, Federal District; CASAN—Indaial, Santa Catarina State; CASAN—Biguaçu, Santa Catarina State; CASAN—Canoinhas, Santa Catarina State; CASAN—Quilombo, Santa Catarina State; CASAN—São Joaquim, Santa Catarina State; CESAN—Espirito Santo State; CIS—Itu, São Paulo State; COMASA—Companhia Águas de Santa Rita S/A, São Paulo State; Companhia Águas de Joinville, Santa Catarina State; COMUSA—Novo Hamburgo, Rio Grande do Sul State; DAE S.A.—Água e EsgotoJundiaí, São Paulo State; DAERP—Ribeirão Preto,

Special characteristics of this learning network: This learning network was planned to be run remotely and nationwide as a follow-up activity to the previous water and energy loss plans. Nevertheless, newcomers got interested and joined this online version of the learning network. The chosen platform for the meetings was "go-to-webinar."

Participant commitment to free online courses is usually lower than to face-to-face ones. Considering this fact, 65 companies were admitted to this online learning network in order to still gather some significant outputs from the learning network.

Main outputs:

- On the documentation level by the participants, the main outputs were:
  - Five satisfying water and energy loss control plans
  - One water loss control plan
  - Two collections proposals of water and energy efficiency projects for the new upcoming administration
  - In depth annual company energy balance and revision of energy bills procedures

---

São Paulo State; DAESBO—Santa Bárbara do Oeste, São Paulo State; DEAGUA—Guaíra, São Paulo State; DAEV—Valinhos, São Paulo State; DESO, Sergipe State; DMAE—Monte Carmelo, Minas Gerais State; EMBASA, Baía State; IGUÁ Saneamento S/A; Prefeitura De Ipeúna, São Paulo State; SAAE—Amparo, São Paulo State; SAAE—Atibaia, São Paulo State; SAAE—Brotas, São Paulo State; SAAE—Carmo Do Cajuru, MG; SAAE—Formiga, Minas Gerais State; SAAE—Guanhães, Minas Gerais State; SAAE—Indaiatuba, São Paulo State; SAAE—Itabira, MG; SAAE—Itabirito, MG; SAAE—Lagoa da Prata, MG; SAAE—Mogi Mirim, São Paulo State; SAAE—Porto Feliz, SP ; SAAE—São Carlos, São Paulo State; SAAE—Sorocaba, São Paulo State; SAAE—Salto, São Paulo State; SABESP, São Paulo State; SAE—Ituiutaba, Minas Gerais State; SAE—Jumirim, São Paulo State; SAEP—Pirassununga, São Paulo State; SAMAE—Blumenau, Santa Catarina State; SAMAE—Brusque, Santa Catarina State; SAMAE—Nova Trento, Santa Catarina State; SAMAE—Pomerode, Santa Catarina State; SAMAE—São João Do Oeste, Santa Catarina State; SAMAE—São Ludgero, Santa Catarina State; SAMAE—Tietê, SP; SAMASA—Três Barras, Santa Catarina State; SANASA—Campinas, São Paulo State; SANEAGO, Goiás State; SANEAR—Rondonópolis, Mato Grosso State; SANEBAVI—Saneamento Basico Vinhedo, São Paulo State; SANESUL, Mato Grosso do Sul State; SAS—Barbacena, Minas Gerais State; SEMAE—Mogi Das Cruzes, São Paulo State; SEMAE—Piracicaba, São Paulo State; SEMAE—São Leopoldo, Rio Grande do Sul State; SEMAE—São José do Cedro, Santa Catarina State; SESAM—Carmópolisde Minas, Minas Gerais State; SSAAP—Cáceres, Mato Grosso State; SAE—Ituiutaba, Minas Gerais State.

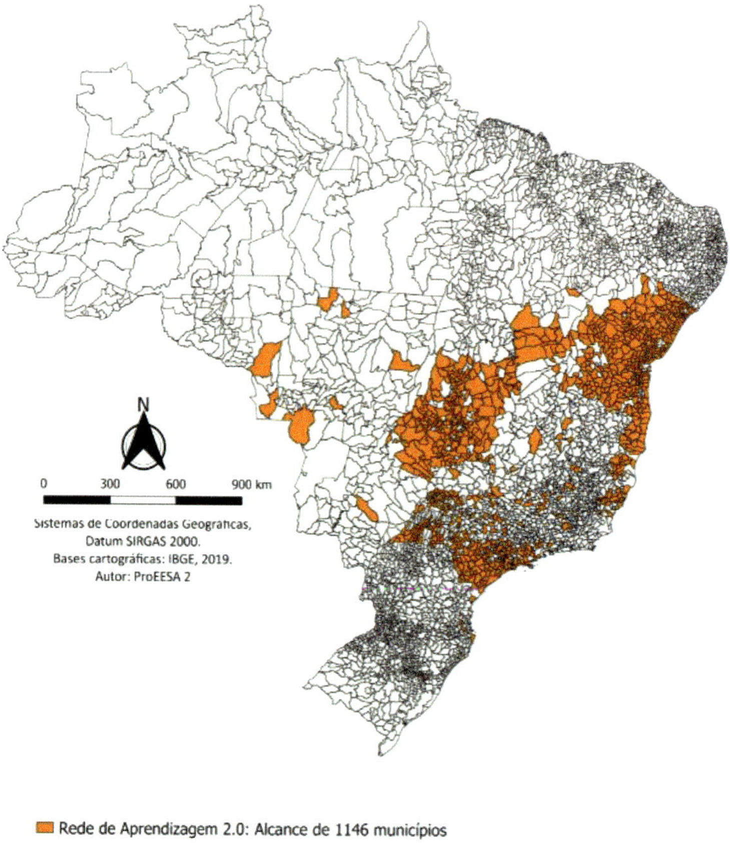

Figure 5.6 Location of the participants and operating area of the participants of the ProEESA 2.0 learning network.

- Amongst major energy efficiency projects with synergies with the learning network, there were:
  - SANESUL, Mato Grosso do Sul State, Implementation of Photovoltaic Energy Production Plant
  - Águas de Joinville, Santa Catarina State, Creation of Department of Water Loss, and Energy Efficiency Management
  - Águas de Joinville, Santa Catarina State, "Smart Water—Optimizing the operation of drinking water supply through the Internet of Things"

- CAESB, Federal District, Implementation of Photovoltaic Energy Production Plant
- SAAE—Indaiatuba, São Paulo State, Constitution of the Water Loss Fight Commission
- Nine learning network participants[16]prepared comprehensive energy efficiency projects at the annual public call of the Energy Efficiency Program (PEE) of the National Agency of Electric Energy (ANEEL). These projects required considerable time and input from technical skilled and qualified teams.

Important side effects: In this learning network, a major shift and challenge was faced by using remote meeting platforms for a wide range of participants while having several levels of maturity amongst participants. Fifty-five videos were recorded and made available on ProEESA's YouTube channel.

5.4.7 ProEESA REG, 2020 Nationwide—Brazil and International—South America

Main goal and execution period: The main goal of this learning network was to provide and discuss incentive tools to regulate and promote energy efficiency. The **ProEESA REG Learning Network on Regulation Instruments for Water and Energy Efficiency** was held between June 2020 and June 2021 (Figure 5.7). Specifically, the learning network focused on regulatory practices such as (i) efficiency benchmarking, (ii) determining the economic level of (real and apparent) water losses, and (iii) data and information auditing through the *ACERTAR methodology*, especially the data related to water, energy, and asset management.

Operational partners and participants: The local operating partner was the Brazilian Association of Regulation Agencies *(ABAR -Associação Brasileira de Agências de Regulação*). The beneficiaries were mainly regulation agencies, responsible for encouraging and

[16]SANEAGO, Goiás State; EMBASA, Baía State; SAE Tietê, São Paulo State; SEMAE—Mogi das Cruzes, São Paulo State; SABESP, São Paulo State; DAESBO—Santa Bárbara DÓ Este, São Paulo State; SAAE—Amparo, São Paulo State; DAE—Valinhos, São Paulo State; Companhia Águas de Joinville, Santa Catarina State.

regulating efficiency, and water operators. This is because they are able to use the same instruments as "internal self-regulation." Water operators also need to be fully prepared for a comprehensive dialog with their regulation agencies. Twenty-four entities[17] with 128 technicians participated from 15 regulatory agencies, five water and sanitation service providers, and four consulting firms in two countries (Brazil and Colombia). Nationwide the initiative covered 13 Brazilian states. These participating entities operate in 1,828 Brazilian municipalities, representing 102 million inhabitants, about 47% of the Brazilian population.

Special characteristics of this learning network: A call by the Ministry of Regional Development was madc to participate in this learning network. The learning network was designed to occur mainly online due to the vast distances between the headquarters of the participating regulation agencies. Although, two face-to-face meetings were planned to follow the quarterly technical meetings of ABAR, due to restrictions associated with the COVID pandemic. These were held online too. Fourteen meetings occurred and 52 videos were recorded and made available on ProEESA's YouTube channel.

[17]Agência Reguladora Intermunicipal de Saneamento do Rio Grande do Sul, (AGESAN-RS), State of Rio Grande do Sul; Agência Intermunicipal de Regulação do Médio Vale do Itajaí, (AGIR), State of Santa Catarina; Agência Nacional de Águas e Saneamento Básico, (ANA), Brazil—nationwide; Agência de Regulação de Serviços Públicos de Santa Catarina, (ARESC-SC), State of Santa Catarina; Agência Reguladora dos Serviços de Saneamento das Bacias dos Rios Piracicaba, Capivari e Jundiaí, (ARES-PCJ), State of São Paulo; Agência Reguladora Intermunicipal de Saneamento, (ARIS), State of Santa Catarina; Agência Reguladora de Serviços de Abastecimento de Água e de Esgotamento Sanitário do Estado de Minas Gerais, (ARSAE-MG), State of Minas Gerais; Agência Reguladora de Saneamento Básico de Natal, (ARSBAN), State of Rio Grande do Norte; Agência Municipal de Regulação dos Serviços Públicos Delegados de Cuiabá, (ARSEC), State of Mato Grosso; Agência Reguladora de Serviços Públicos do Estado de São Paulo, (ARSESP), State of São Paulo; Agência Tocantinense de Regulação, Controle e Fiscalização de Serviços Públicos, (ATR), State of Tocantins; Agência Tocantinense de Saneamento, (ATS), State of Tocantins; BEU Engenharia, N.A.; BRK Ambiental, N.A.; Companhia de Saneamento de Alagoas, (CASAL), State of AL; Comisión de Regulación de Agua Potable y Saneamiento Básico, (CRA), Colombia; Companhia Águas de Joinville, (CAJ), State of Santa Catarina; DeodeInovação e Eficiência; Higra Industrial LTDA.; KMR Energia e Meio Ambiente Ltda.; Companhia de Saneamento do Paraná, (SANEPAR), State of Paraná.

Figure 5.7 Location of the participants and influence area of the participants of the ProEESA_REG learning network.

Main outputs: At the end of the network, 55 tasks were successfully completed by participants. Among the 55 concluded tasks were:

- Planning an audit on data related to water energy and asset management, as a regulator or as an operator internal auditor—according to its specific regulation[18]

[18]*Portarianº 719, de 12 de dezembro de 2018 of the Ministry of Regional Development.*

- Executed audit—Work papers filled regarding the energy consumption monitoring process
- Executed audit—Work papers filled regarding macro and operational water metering (operational process)
- Executed audit—Work papers filed out of the commercial process (client database, contracts, and water metering)
- Executed audit—Work papers filled regarding the asset management process
- Executed audit—Work papers filed out of the network cadaster process
- Calculation of energy efficiency in pumping stations—Use of the CEN—normalized energy consumption indicator (kWh/$m^3 \times 100$)
- Simulation of energy efficiency auditing in pumping stations with the CEN indicator (kWh/$m^3 \times 100$)
- Applying and balancing fines and penalties in situations of low efficiencies CEN indicator (kWh/$m^3 \times 100$)
- Calculation of the economic level of under-measurement of water meters—apparent losses
- Calculation of the economic level of frauds—apparent losses
- Calculation of the economic level of physical water losses—real losses
- Data Envelope Analysis (DEA) and Data Frontier Analysis (DFA) for efficiency
- Asset management based on average network age—calculation of adequate network replacement

Important side effects: Brazil is going through a major water and sanitation sector reform[19] that has been discussed for many years and finally set in motion in July 2020 when the according framework legislation passed. Follow-up regulation is being elaborated (2021–2023) including more than 20 new norms by the national water regulation agency ANA. ANA is conducting high-quality public discussion based on conceptual drafts within three years (2021–2023) and the ProEESA Reg Learning network contributed for an improved public discussion in the aspects of asset management, water losses, and energy efficiency.

[19]*Novo Marco Regulatório do Saneamento" (Lei Federal 14.026, de 15 de julho de 2020).*

5.5 Some Approaches to Increase Commitment

Some approaches were used to keep participants committed and motivated over a large period of time. A few are described below.

1. **Call for participation—selection criteria**
   The (face-to-face) learning network needed a limited number and optimum size of participants for engagement. Up to 15 organizations with 2–3 persons each were considered a reasonable size. Organizations were mainly selected by call, and preference was given to the most engaged ones.
2. **Midterm and final evaluation**
   A midterm evaluation report was issued by the network coordinator, pointing out which tasks were concluded, and which needed to be concluded by the participating beneficiary organization. The evaluation report was sent to the participant's supervisor for information. With the midterm evaluation, the participants were reminded that this learning network was not passive training, but tasks and exercises needed to transpose to the company's setting. Also, a final evaluation was issued.
3. **Handing out products to supervisors and mayors at a special closure event of the learning network**
   At the closing session of the events, participants handed out their finished product outputs and the water and energy management plans to their supervisors or even the city mayor. In this way, the mayor acknowledged the team's commitment, increased his support for the proposals, and identified efficient projects in the plan.
4. **Evaluation of the execution of the initial intentions**
   Later, in the online learning networks, the number of participants increased while having diversified interests and priorities. To set clear individual goals, participants were asked to indicate during the selection process the topics that were on priorities. The midterm's participant's evaluation compared their progress regarding their initial intentions. Some of the options for the ProEESA 2.0 and ProEESA REG Learning Network were:

- Revision, expansion, or monitoring of the existing water and energy management plan
- Proposals and energy efficiency project elaboration for the next administration (after elections)
- Financing preparation of energy efficiency projects (though several possibilities)
- Implementing a full or a partial ACERTAR (internal or external) audit on water, energy, and asset management data
- Calculation of pumping efficiency using the CEN Indicator—PH5 (kWh/m^3 × 100)
- Calculation of the economic level of apparent (under measurement) water losses, frauds, and real losses
- Sunshine regulation—analysis of historical series of water/energy losses
- Asset regulation and management
- Average age of water supply networks and losses

5.6 Publications Associated with Learning Networks in Water and Energy Efficiency

Throughout the learning networks, topics have been prepared, either to support participants in the elaboration of their outputs or for holding important findings in discussions (mentioned below in publications to support municipal water operators). The purpose of other publications aimed at supporting partner organizations and the relevance of some publications spilled beyond the framework of the learning networks and could be considered sector thematic milestones. The main ones are hereafter mentioned.

5.6.1 Publications to Support Municipal Water Operators

1. **Guidelines for Water and Energy Loss Management Plans**
   This publication[20] includes the (i) Template of the Water and

[20]Portuguese title: *"Linhas orientadoras para elaboração e acompanhamento de Planos de Gestão de Água e Energia em Sistemas de Abastecimento de Água"* AKUT Umweltschutz Ingenieure Burkard und Partner, Rita Cavaleiro de Ferreira, André

Energy Loss Management Plans and (ii) Guidelines to help customize and fill out the Template of the Water and Energy Loss Management Plans. It was the foundation for the outputs of the municipal learning networks. The template of the plan was derived from a version prepared by the Portuguese National Laboratory for Civil Engineering (LNEC) for Portuguese service providers and was adapted to the Brazilian context. For the Brazilian learning networks, a guideline for filling out and interpreting the template was produced.

2. **Guidelines for Calculating Standardized Pumping Performance and Saving Potential**
   This publication[21] helps users to calculate standardized pumping performance through the indicator (kWh/m^3 × 100) and calculate its energy-saving potential. It offers approaches to deal with when data is inaccurate or insufficient and still extracts useful information on energy efficiency. The publication has also an Excel spreadsheet for direct calculation.

5.6.2 Publications to Support Partner Organizations

1. **ACERTAR auditing work papers on water, energy, and asset management data**
   This publication/workpapers[22] are control tests to help regulators audit water, energy, and asset management data of water and sanitation service providers. It is based on the ACERTAR methodology (http://www.acertarbrasil.com/), which covers topics more than water and energy. ProEESA joined efforts with ACERTAR because the information

---

Lopes de Oliveira, Brasilia-*Brazil* February 2018 –Available at https://www.gov.br/mdr/pt-br/assuntos/saneamento/proeesa/redes-de-aprendizagem-1 select "Templates para o plano de gestão de água e energia.

[21]Portuguese title: *Cálculo do Consumo Energético Normalizado e do Potencial de Economia em Instalações de Bombeamento*"AKUT Umweltschutz Ingenieure Burkard und Partner, Rita Cavaleiro de Ferreira, André Lopes de Oliveira, Daniel Manzi, João Geraldo Ferreira Neto, André Galvão Silveira, Jessica Rocha Gama, Brasilia-*Brazil,* October 2017. Available at (there are two files so it is necessary to pick this publication): https://antigo.mdr.gov.br/saneamento/proeesa/biblioteca-virtual/4441-publicacoes-proeesa). It was available 2018—2021 at https://antigo.mdr.gov.br/saneamento/proeesa/biblioteca-virtual/4441-publicacoes-proeesa; currentynotontheMinistry site.

[22]Currently available at http://www.acertarbrasil.com/para-agencias-reguladoras/.

management it addresses is also relevant to water and energy efficiency. These important topics were included in the learning network editions after 2019 for both regulation agencies and water service providers.

2. **ACERTAR information and indicator credibility calculation**

   All the data on water, energy, and beyond is used to indicate accuracy. Their accuracy and credibility depend on the reliability of information management processes. This "publication/spreadsheet"[23] calculates the grade given to the credibility of data and indicators.

3. **Guidelines for implementing Learning Networks in Water and Sanitation Services—Tools to accelerate good practices implementation and propagation of knowledge transfer**

   This publication[24] aimed at regional organizations interested in setting up their learning networks and accelerating the spread out of specific know-how and best practices. It contains important steps and tips for organizing and conducting learning networks. It also includes risks of failure of the learning networks and how to minimize and mitigate those risks.

4. **Proposals to address Energy efficiency in National Water and Sanitation Sector Norms (Reference Standards)**

   This publication[25] gathered several tools at the regulatory level to promote energy efficiency in water and sanitation service providers. Tools were discussed (and partially tested) during the learning network with the regulation agencies.

---

[23]Currently available at http://www.acertarbrasil.com/para-agencias-reguladoras/.

[24]Portuguese title: "*Roteiro para constituir Redes de Aprendizagem com prestadores de serviços de saneamento- Mecanismos de aceleração na adesão a boas práticas e propagação de conhecimento*" AKUT Umweltschutz Ingenieure Burkard und Partner, Jonas Gonçalves, Rita Cavaleiro de Ferreira, Thalita Salgado Fagundes, Ricardo Hübner, Jessica Rocha Gama, Brasilia April 2018. Currently Available at https://antigo.mdr.gov.br/images/stories/ArquivosSNSA/proeesa/pdf/roteiroredede aprendizagembrasil.pdf.

[25]Portuguese title: "Propostas para abordar a eficiência energética em Normas de Referência Nacionais de Saneamento" AKUT Umweltschutz Ingenieure Burkard und Partner, Rita Cavaleiro de Ferreira et al., Brasiliajul2021—Currently available at https://www.gov.br/mdr/pt-br/assuntos/saneamento/proeesa/EficienciaEnergeticaNormasRegulatoriasSaneamento.pdf.

It includes proposals for 11 out of the 22 norms that are currently in elaboration in the period (2021–2023) for nationwide application.

5.6.3 Publications to Address the Sector

1. **Performance Monitoring and Verification for Typical Water and Sanitation Projects**

   This publication[26] docked on the Energy Efficiency Program of Programme (PEE) from the Brazilian Electricity Regulatory Agency (ANEEL). Measures for energy efficiency in the Water and sanitation sector were theoretically eligible in the programme, but not due to knowledge and institutional barriers. In the first years, the programme attracted efficiency projects in the areas of lightning, refrigeration, and motors. ProEESA tried to close the knowledge gap in the water sector by first transferring the necessary cost-benefit calculations and measurements of efficiency projects to specific water-related projects. Afterward, it prepared templates for all typical energy efficiency projects in the water sector.

2. **Water losses—Guide to determining the economic level and progressive targets**

   This publication[27] was meant to help municipalities, service providers, concessionaires, and regulation agencies to set reasonable goals for water losses while considering the context of water availability, scarcity, production, and infrastructure costs. Economic levels of water losses vary across the globe and in such diverse countries like Brazil.

   It was prepared in 2020 and 2021 and received important contributions from the participants of both learning

[26]Portuguese title: " "Projetos de Eficiência Energética em Saneamento—Guia auxiliar para os Procedimentos do Programa de Eficiência Energética—PROPEE da ANEEL" AKUT Umweltschutz Ingenieure Burkard und Partner, Rita Cavaleiro de Ferreira et al., Brasilia Abril2019—Available at https://antigo.mdr.gov.br/images/stories/ArquivosSNSA/proeesa/pdf/2019-05-21_Guia_Saneamento_PROPEE_ANEEL_V2.pdf.

[27]Portuguese title: "*PERDAS DE ÁGUA: Guia para determinar o nível econômico e metas progressivas de controle*" AKUT Umweltschutz Ingenieure Burkard und Partner, Wyatt A; Rita Cavaleiro de Ferreira et al., Brasilia March2021—https://antigo.mdr.gov.br/images/20210325_Nivel_economico_de_perdas_de_agua__Brasil.pdf. Access to spreadsheet https://drive.google.com/file/d/1y7MYWI_Vor4xvcDJ_ud0ToOQv9tSPJ45/view.

networks (ProEESA 2.0 online and ProEESAREG). Note that the methodology described in this document has already been adopted by the regulation agency of the São Paulo State (ARSESP)[28] in determining the regulatory goal of losses for the third ordinary tariff review of the estate water service provider SABESP with validity and application in the years 2021–2024. Also, the regulation agency AGESAN in Rio Grande do Sul Agency applied this methodology in 2021 and was awarded a prize for its best practices in regulation by the University of Law FGV—Rio.[29]

3. **Guidelines (and script) to extract digital data of the electricity bills for automated analysis and operational recommendations**

   The fact that energy bills are issued by energy companies to water utilities in paper form or PDF is a huge barrier to energy efficiency monitoring especially when utilities handle hundreds and thousands of electricity bills every month. This publication[30] and script extract digitally the data and make past and trend analyses for energy consumption and several other aspects that need to be actively monitored and reacted to when deviations happen.

[28]The whole documentation can be accessed on http://www.arsesp.sp.gov.br/SitePages/consultas-publicas.aspx, using "Consulta pública nº:11/2020".

[29]https://abar.org.br/arsesp-e-agesan-vencem-premio-fgv-de-melhores-praticas-em-regulacao/.

[30]Portuguese title: "Manual de instruções e recomendações para rotina de digitalização de faturas de energía elétrica da CPFL e da Elektro" AKUT Umweltschutz Ingenieure Burkard und Partner, Fernando Finger, Daniel Hoffmann e Kalffman Schuch, Rita Cavaleiro de Ferreira, Brasilia novembro 2021—It was available in 2021 at https://antigo.mdr.gov.br/saneamento/proeesa/biblioteca-virtual/4441-publicacoes-proeesa; currently not on the Ministry site.

Chapter 6

Other Latin American Experiences

Javier Hueichapán Henríquez,[a] Luis Enrique Manzano Villafuerte,[b] Angel Alfredo Núñez Barretoa,[c] Welner Collejo Jerez,[c] Claudio Carpio,[d] and Jairo Fernando Calvo[e]

[a]*Empírica Consultores, Chile*
[b]*Universidad de las Fuerzas Armadas (ESPE), Ecuador*
[c]*National Office for the Control of the Rational Use of Energy (ONURE), Matanzas, Cuba*
[d]*Fundación Bariloche, Argentina*
[e]*Independent consultant on energy efficiency, Nicaragua*
javier.hueichapan@gmail.com, seifercalvo@gmail.com

As mentioned before, the Latin American region initiated the path for the development of LEENs recently. To incorporate different experiences, this chapter includes more recent or novel experiences in the deployment of LEENs, such as the cases of Chile, Cuba, Ecuador, Colombia, and Nicaragua.

The methodological approach followed by these countries is, to some extent, different from one of the rest of the chapters, as some

Learning Energy Efficiency Networks: Applications and Challenges in Latin American Countries
Edited by Claudio Carpio and Marina Yesica Recalde
Copyright © 2026 Jenny Stanford Publishing Pte. Ltd.
ISBN 978-981-4968-82-9 (Hardcover), 978-1-003-77468-6 (eBook)
www.jennystanford.com

DOI: 10.1201/9781003774686-7

of the cases are developed by experts as authors, while others are the result of interviews with in-country experts.

6.1 Chilean Energy Efficiency Networks in the Mining Sector[1]

Javier Hueichapán Henríquez
Empírica Consultores, Jefe de proyectos de Sustentabilidad, Chile
jhueichapan@empiricaconsultores.cl, javier.hueichapan@gmail.com

6.1.1 Introduction

The GIZ Gmbh within the framework of its Programme **"Programa 4e Chile"** (Renewable Energy and Energy Efficiency Program), supports the Government of Chile in activities aimed at promoting the reduction of CO_2 emissions, with research and development, advice for improvements to the regulatory framework, activities aimed at promoting business and project financing, and to improve the sustainability of the energy sector in Chile.

The project *Raw Materials and Climate* (MaPriC for its acronym in Spanish) funded by the Federal Ministry of the Environment, Conservation of Nature, Nuclear Safety, and Consumer Protection (BMUV) supports the Ministerio de Energía (Ministry of Energy) and the Agencia de Sostenibilidad Energética (Energy Sustainability Agency) in the generation, replication, and adaptation of energy efficiency initiatives, which among its activities aims to accelerate the introduction in the country of concrete measures for increasing energy efficiency and reducing GHG emissions. One of the main activities of the MaPric was the creation of the first Learning Network on Energy Efficiency and Emissions Reduction in Mining, which was made by 16 mining companies from large, medium, and small sectors mining in Chile, generating various studies and learning for this sector[2].

---

[1]Based on *"GIZ—Memoria Red de Aprendizaje—Implementación de la primera red de Aprendizaje en Minería—Chile"* March 2023.
[2]More information at this link: https://4echile.cl/.

The first LEEN was created in March 2021 in Energy Efficiency and Emission Reduction in Mining in Chile, a part of GIZ, and with the sponsorship of the Ministerio de Energía and the Agencia de Sostenibilidad Energética, association Sociedad Nacional de Minería (SONAMI, National Mining Society), association Consejo Minero and Corporation Alta Ley. This network was based on the LEEN methodology for EMS. It actively worked with professionals from 16 mining companies to share technical knowledge and collaborative learning and advance in the implementation of projects related to the themes of the network by the participating companies.

In 2023, the execution of the "Team Europe Project for the Development of Renewable Hydrogen in Chile" began, which is implemented by GIZ and the Spanish Agency for International Development Cooperation (AECID), co-financed by the European Union and the Federal Ministry of Economy and Climate Protection of Germany (BMWK), being the main vehicle to implement the Team Europe initiative for the Development of Renewable Hydrogen in Chile, a joint effort of the European Union and its Member States, to foster cooperation with Chile in the development of its renewable hydrogen economy.

This initiative seeks to replicate the successful experience of the Mining Network, with the Latin American Learning Network on Renewable Hydrogen, which is already made up of 24 companies and institutions from the entire hydrogen value chain.

As follows, this chapter describes the following points. Firstly, a general introduction to the topic of the LEEN created in Chile, with the background, the description of the general objectives, and a general summary of the actors that made it up. Secondly, a description of the working procedures of a LEEN, its formation to the evaluation of the results, and closure. Thirdly, an overview of workshop activities, emphasizing benefits, achievements, and learning from the network. Finally, the lessons learned in the framework of the Learning Network are described and include general conclusions and recommendations.

6.1.2 General Description of the Chilean LEENs

Since the end of 2019, the EEM project has sought to generate the instances to implement the first Learning Network in Mining in Chile. Due to the limitations of face-to-face meetings caused by the global crisis generated by the COVID-19 pandemic, the first Learning Network in the sector was initiated in March 2021.

It was possible to develop continuous work during the first year of the network and even generate the extension of this to a second year at the request of the companies participating. With this, the first stage guided by GIZ successfully ended.

By implementing the LEEN methodology in Chile, GIZ retained part of the working methodology of the learning network; however, it had to be adapted to the Chilean reality, local conditions, and times of the mining sector, and the global context concerning the COVID-19 pandemic. For this reason, the workshops were held during the two years online. The meetings of the network were held from March to November 2021–2022, only except for the face-to-face closing meeting of the first two years of the network held in November 2022. By modifying the methodology, it was possible to advance in obtaining results of collaborative work, connecting, and generating paradigm shifts such as counting with a trusted space to share knowledge among its participants.

The first year of the Learning Network in the mining sector (2021) was successful from different views, expressed by the initial 14 participating companies when showing their interest and commitment to continue with the network in the second stage. It shows that the participants recognize the added value of working in the network and in technical topics that were of interest to, recognizing and valuing at the same time, GIZ's coordination work in achieving the prioritization of topics together with network participants.

The second year of the network (2022) kept the 14 companies and added two additional companies, totaling 16 companies from the mining industry. It should be noted that the network was made up of large, medium, and small mining companies.

Based on the experience gained during the first year, the experience was made a little more ambitious in the second year. This Learning Network was undoubtedly successful in aspects such as active participation and the generation of technical collaboration. It also left many lessons for both the actors directly involved with the mining sector and all those institutions that care about the implementation of energy efficiency measures in the mining and energy industry, in particular.

Most network participants appreciate a change in their thinking, as they recognize the benefits and opportunities involved in participating in this type of instance. The Learning Network is perceived as a means of allowing collaborative actions to be carried out in the mining industry with common objectives.

The second year of the Mining Learning Network in Chile ended with a face-to-face activity in November 2022. In that activity the companies had the opportunity to meet again, get to know each other personally, share impressions, learn throughout the period, and deliver as a group the vision of how to continue with this work.

In the documentation aspect, the network worked and generated five studies, which delivered results for each company, as well as overall results. Regarding these common results, the companies on the network reached a full consensus on the value of sharing them with the mining sector. For this reason, the studies published by the network were:

- "Study Emissions in the Mining Sector 2020"
- "Energy Performance Indicators of the Mining Sector in Chile"
- "Guide for Generating Electro-Mobility Roadmap in Mining"
- "Energy Efficiency Guide to Ore Milling"
- "Study Emissions in the Mining Sector 2021"

The work of the network is expected to continue from 2023 onwards due to the high interest of the participant mining companies in continuing with the collaborative work of the network. There are other mining companies interested in joining this initiative.

6.1.3 Context of the Mining Learning Network in Chile

GIZ's project "Energy Efficiency in Mining, Raw Materials, and Climate" of the "Programa 4e Chile" (Renewable Energy and Energy Efficiency Program), started in mid-2019 and one of its activities was to provide technical support to mining companies in promoting energy efficiency with the ultimate goal that they could reduce their greenhouse gas emissions associated to energy consumption.

Prior to the official creation of the network in March 2021, GIZ worked on raising awareness and making the methodology, advantages, and benefits of the Learning Networks known to the mining sector actors. This was achieved through bilateral meetings with the public sector (Ministerio de Energía and the Agencia de Sostenibilidad Energética and COCHILCO, Chilean Copper Commission) and private sector (Consejo Minero and SONAMI).

As a continuation of this first stage, two instances of work were carried out with the mining sector to invite them to participate actively in the creation of the network. In addition, a summary document of the previous experiences in Learning Networks was generated at the Latin American level. The background of the methodology was also included in the document.

The first meeting, which was called "Informative Event" and held on November 28, 2019, was attended by 13 large, medium, and small mining companies in Chile, together with the associations of SONAMI and Consejo Minero, and Ministerio de Energía, Agencia de Sostenibilidad Energética, Corporation Alta Ley and COCHILCO. The Learning Network methodology was reported, worries were collected, exclusions (topics not to be touched on the net such as issues that might affect free competition, etc.), barriers, and opportunities to participate in a LEEN.

The second working meeting, which was held on January 15, 2020, was attended by 8 large and medium-sized mining companies in Chile, together with the associations of SONAMI and the Consejo Minero, the Ministerio de Energía, the Agencia de Sostenibilidad Energética, the Corporation Alta Ley and COCHILCO. Topics of interest were collected and guidelines for the future network. In addition, voting and prioritizing of technical topics of interest proposed was carried out by mining companies, classifying them into main topics and secondary ones.

The main topics prioritized by the companies interested in forming the network were:

1. Use or replacement of fuel for energy saving (e.g., hydrogen, synthetic fuels, renewable energies such as geothermal, solar energy, etc.)

2. Energy efficiency in mineral transport and electro-mobility
3. Energy efficiency in ore grinding
4. Energy efficiency in the heating process

The cross-cutting themes proposed by the companies interested in forming the network were:

1. Traceability and certification
2. Measurement, reporting, and verification (MRV)
3. Training
4. Governance
5. Good practices
6. Innovation

It was agreed that these issues would be included as priorities in the network's work when forming.

According to the planning and given the interest of the mining companies, the network considered starting its activities in April 2020, but its official start was affected by the global crisis generated by the pandemic. The mining companies interested in participating (more than 13) had to focus their efforts on maintaining production and taking the utmost care of the health of their workers.

That is why, in the following months of 2020, GIZ's EEM project focused on finding the best options to continue with the work of the network. In parallel, EEM project activities were carried out, such as opening applications to receive technical support for energy efficiency projects in mining and conducting online technical seminars (webinars).

At the same time, the COVID-19 pandemic was advancing in terms of the management of information on its prevention and the operation of the mining companies was normalizing in terms of protocols. By the end of 2020, GIZ's EEM project decided to carry out two simulation workshops of the network to involve companies that were interested in entering it.

The first one (September 2020) was a technical session on ore milling. The second theme offered to the network was to measure the carbon footprint in the mining sector for each company participating (November 2020). In this session, GIZ proposed the name of the network: "Energy Efficiency and Reduction Emissions Network". It was very well-received, so it was maintained. In both sessions, the methodology of learning networks was simulated, to motivate the participation of companies and exemplify the type of workshops that would be held.

In parallel, two training courses were held for professionals who could accompany a Learning Network, bringing the experience of other networks implemented worldwide.

The first training (face-to-face modality, November 2019) was for energy efficiency advisors so that they could learn the methodology and become future Technicians Companions for the Learning Network. The second training (online modality, November 2020) was for professionals in the field of coaching to become future Learning Network Moderators.

Another relevant precedent was that, in December 2020, the large-scale mining companies belonging to the Consejo Minero committed to accounting for and reducing their emissions. These large copper mining companies account for 97% of national copper production and voluntarily committed targets to reduce greenhouse gas emissions (GHG).

It was the first productive sector to assume this type of commitment, which confirmed the interest of mining in contributing to the fight against climate change and more specifically to compliance with Chile's objectives of carbon neutrality by 2050 and emission reductions by 2030, set in the Nationally Determined Contributions (NDCs). This commitment was also consistent with the principles of climate change that the Consejo Minero made public in 2015 and updated in 2019.

In February 2021, the Energy Efficiency Law (No. 21.305/2021) was enacted, which aims to promote the use of rational and efficient energy resources. Among the contents of the Law is to promote the management of energy in the country's large consumers, to promote and eliminate barriers to energy efficiency, and thus increase the competitiveness and productivity of companies of the country.

In this context, the Law, in Article 2, establishes a general obligation for large companies (with sales over 100,000 UF/year) and those with higher consumption to 50 Tcal, defined as Consumers with Energy Management Capacity (CCGE), to report their energy consumption and energy intensity and to actively manage their energy by implementing and maintaining an energy management system (EMS).

The regulation of Law (N°21.305/2021) established that large energy consumers must actively manage their energy consumption. Specifically, Consumers with Energy Management Capacity (consumption) over 50 Tcal/year) must implement an energy management system (EMS) and must report annually to the Ministerio de Energía their energy consumption and other indicators, which will be annually prepared as a public report.

The Energy Efficiency and Emission Reduction Network was formally launched on March 23, 2021, with the launch headed by the Minister of Energy and Mining, Juan Carlos Jobet. The network was made up of 14 companies and had its first official session on April 13, 2021.

One of the definitions that the network took from its beginning was the confidential handling of information that would be discussed in the sessions. In these working sessions only the companies, together with the organizing entity (GIZ), would participate.

During the first two years of the LEEN, 16 small, medium, and large mining companies participated.

The companies were: 1. Anglo American Chile; 2. Antofagasta Minerals; 3. CEMIN Mining Holding; 4. Cerro Negro Mining Company; 5. Codelco; 6. National Mining Company (ENAMI); 7. KGHM Chile; 8. Lundin Mining Candelaria Mining District; 9. Minera Cruz; 10. Minera Rafaela; 11. Minera San Pedro; 12. Minera Tres Valles; 13. Punta del Cobre Society (PUCOBRE); 14. MCS; 15. Teck Resources Chile and 16. Minera Valle Central

GIZ's project "Energy Efficiency in Mining, Raw Materials, and Climate" came to an end in February 2023 having left learned and motivated professionals to continue working in a learning network.

6.1.4 Lessons Learnt in Chile

The implementation of the first LEEN in Chile intended to serve as a reference for the continuity of the mining network and the creation of future new learning networks in other sectors or productive areas. It sought to share good practices, the experience, and the learning of the participating companies.

The Learning Network of the 4e Renewable Energy and Energy Efficiency program, which GIZ developed to improve energy performance and reduce emissions for mining in Chile, proved to be a great challenge from opening to closing.

When it comes to the mining industry and the context in which its activities were developed on the start date of the network (the global COVID-19 pandemic), the challenge was extremely different. So, the methodology of the Learning Networks had to be adapted to the existing mining and national reality in such a way that this network could be initiated to obtain the benefits of the network and also achieve the expected results of the methodology.

Companies have found the information shared, analyzed, synthesized, and delivered valuable and believe that the network has allowed them to work on their energy management systems and improve their processes.

Prior to their participation in the GDR, some companies had few elements to build their energy performance indicators (EDIs). There were some gaps in knowledge to efficiently manage milling processes, how to measure and criteria to use in the carbon footprint, and how to move forward in building a roadmap to address the challenges of electro mobility. During the development of the LEEN, companies worked on setting objectives and targets, based on a common objective.

The representatives of the participating companies, having both technical profiles and experience in management systems, generated an important synergy that was positively used in the activities of the LEEN.

Fostering a cordial and trusting atmosphere among the LEEN participants was of great relevance, as it produces a sense of belonging, which is crucial for creating ties between them. These links based on mutual trust will allow the exchange of information and experiences.

Although the workshops began exclusively with technical topics, an additional 15 minutes were spent prior to each session, which sought to give participants the opportunity to get to know each other better and to create links between them.

In the first year of the network, 14 companies initially participated, and then in the second year, two additional companies joined. Many companies expressed interest in continuing to participate actively in the LEEN after the end of the first year and after the end of the second year. Although on average 64% participation was obtained by the companies, attendance in the workshops was always a major challenge.

In this Learning Network, technical companions from five consulting companies collaborated with different thematic focuses depending on the technology or theme required to address. It is highlighted that it is very important that the technical accompaniment knows in advance the methodology of the LEEN that will be used in the workshops and that they naturally make it their own, to generate the results in terms of participation of the attendees and to share information among them. The network's workshops should not be seen as a seminar to go only to listen, but to participate actively.

The methodology implemented and adapted from the Learning Network is a very valuable tool to help achieve goals and support companies in improving their energy performance. The companies' commitment and willingness were key to obtaining successful results. These are documented as a testimony of the efforts made by both GIZ and the different participating companies and as a reference for the creation of new networks in Chile.

On average, 55% of participation was obtained from large-scale mining companies, 40% from medium-sized mining companies, and an average of 2% from small-scale mining companies.

This project brought together very different companies from the point of view of size and sector, but with common interests in terms of the will to improve energy performance and with diverse needs. All these companies run into similar problems when implementing energy efficiency projects.

Some of the barriers identified are that some companies may expect the solution to their problems to be found outside the group set of the network. Also, the sector and professionals may be reluctant to leave their comfort zone and not want to share knowledge.

This network is the first instance at the national level where the participation and commitment of mining companies has been achieved for 2 years, actively participating in generating public studies. Through collaborative work, they have managed to achieve the defined goal and are motivated to continue with greater and greater challenges.

Although good participation was achieved during the networking sessions, more effort is required to get more companies to participate and want to share knowledge with other companies.

The scope of the network allowed GIZ to disseminate more directly to the network other activities to support the mining sector that were being carried out within the framework of the EEM project (training courses given by the SE Agency, Industrial Energy Manager Diploma mining version, available technical support, online seminars, web platform Energy in Mining, etc.). At the same time, in addition to sharing within the network, companies publicly showed their projects in webinars organized by GIZ's EEM project.

For the initiators of the LEEN, this experience has exceeded their expectations. Analyzing what has been learned in this first LEEN—the conclusion is that this methodology, adapted to the Chilean reality and implemented for the first time worldwide in the mining industry, has managed to achieve the objective set by GIZ and generate a great impact within the mining ecosystem.

6.2 The Ecuadorian Case

Luis Enrique Manzano Villafuerte
Universidad de las Fuerzas Armadas (ESPE), Ecuador
luis.manzanovillafuerte@gmail.com, lemanzano1@espe.edu.ec

6.2.1 Context Situation: Origin of the Ecuadorian LEEN

Ecuador ratified its commitment to contribute to combating climate change through the implementation of its Nationally Determined Contributions (NDCs), which include various actions aimed at reducing GHG emissions and that are aligned with the country's national and sectoral policies within which the energy sector is considered to obtain a greater impact.

In this context, the flagship program of the European Union called EUROCLIMA+ (EC+) promotes environmental sustainability and fights against climate change in Latin America. Its aim is to reduce its impact and effects in the region by promoting mitigation and adaptation to climate change, resilience, and investment. EUROCLIMA+ develops actions that are considered of strategic importance for the implementation and/or updating of the NDCs in its 18 partner countries.

In 2019, the Government of Ecuador, through the Ministry of Environment (MAE), which has now merged with the Ministry of Energy and Mines, and the new ministry is called the Ministry of Environment and Energy. Initiated the process known as the country dialog with EUROCLIMA+, to identify priorities to support the implementation of the NDC identified by the Inter-institutional Committee on Climate Change (CICC), within the framework of the preparation of the country's first NDC. Ecuador submitted its NDC to the United Nations Framework Convention on Climate Change (UNFCCC) on March 26, 2019.

As a result of this dialog process, a series of priorities were identified to request EUROCLIMA+ support, which the program subsequently analyzed based on its criteria and response capacity. After this process, the program considered it pertinent to provide technical assistance support to Ecuador in the implementation of NDCs in the energy sector, specifically to the Ministry of Energy and Mines of Ecuador currently Ministry of Environment and Energy.

6.2.2 The Legal and Regulatory Framework to Boost Energy Efficiency in Ecuador

Article 15 of the Constitution of the Republic determines that the state shall promote, in the public and private sectors, the use of environmentally clean technologies and non-polluting and low-impact alternative energies.

Article 413 of the Constitution of the Republic of Ecuador provides that the state shall promote energy efficiency, the development and use of environmentally clean and healthy practices and technologies, diversified low-impact renewable energies that do not jeopardize food sovereignty, the ecological balance of ecosystems, and water rights.

Article 414 of the Constitution of the Republic of Ecuador establishes that the state shall adopt adequate and transversal measures to mitigate climate change, by limiting greenhouse gas emissions, deforestation, and atmospheric pollution; take measures to conserve forests and vegetation; and protect populations at risk.

Organic Law of the Public Service of Electric Energy (LOSPEE), in the article attributes to the Ministry of Electricity and Renewable Energy (now Ministry of Energy and Mines—MEM) according to Executive Decrees 399 of May 15, 2018 and 400 of April 14, 2022); the stewardship and planning of the electricity sector; and, in the area of energy efficiency, it is responsible for defining and implementing the mechanisms to achieve it, by the provisions of the Constitution and the Law.

Article 13 of the LOSPEE states that the MEM "shall be responsible for the planning of the electricity sector, renewable energies, and energy efficiency, by the provisions of the Constitution of the Republic, the National Development Plan, and the national policy issued by the President of the Republic". For this, it attributes the elaboration of the National Energy Efficiency Plan (PLANEE), in coordination with the Secretariats of State and institutions whose functions are related to the use of energy, which will be mandatory for the public sector and indicative for the private sector.

Article 74 of the LOSPEE states that energy efficiency will have, as a general objective, obtaining the same service or product with the lowest energy consumption. In particular, the following:

- ✓ Promote efficiency in the economy and society in general and in the electricity system.
- ✓ Promote values and behaviors aimed at the rational use of energy resources, prioritizing the use of renewable energies.
- ✓ Promote the rational use of electricity by consumers or end users.
- ✓ Encourage the reduction of production costs through the efficient use of energy, to promote competitiveness.
- ✓ Reduce the consumption of fossil fuels.
- ✓ Guide and defend the rights of the consumer or end user.
- ✓ Experience in Ecuador on Learning Networks in Energy Management Systems.
- ✓ Reduce environmental impacts with the sustainable management of the energy system.

Article 75 of the LOSPEE attributes to the MEM the task of establishing energy efficiency policies and standards that seek greater efficiency in the use of energy sources by consumers or end users, in accordance with the National Development Plan.

Article 76 of the LOSPEE cites that the state, through the Ministry of Electricity and Renewable Energy, will promote energy efficiency through incentives or punishments, which will be defined in the general regulations of this law and the corresponding regulations.

The PLANEE 2016–2035, published on May 18, 2017, aims to "Increase the efficient use of energy resources through the implementation of energy efficiency programs and projects in sectors related to energy supply and demand, reduce the import of petroleum products, contribute to the mitigation of climate change, and create a culture of energy efficiency supported by a solid legal and institutional base."

The Organic Law of Energy Efficiency (LOEE) was published through Supplement to Official Registry No. 449 of 19 March 2019; and aims to "establish the legal framework and operating regime of the National Energy Efficiency System (SNEE), and promote the efficient, rational, and sustainable use of energy in all its forms, to increase the country's energy security (...)"

In Article 2, the LOEE declares the efficient, rational, and sustainable use of energy, in all its forms, of national interest and as state policy as a key element in the development of a solidarity society that is competitive in production and concerned about economic and environmental sustainability.

Article 29 of the Law of the Ecuadorian Quality System (LSEC) provides that "technical regulations include the preparation, adoption, and application of technical regulations necessary to safeguard the objectives related to safety, the health of human, animal and plant life, the preservation of the environment, and the protection of the consumer against deceptive practices."

Official Register No. 528 of July 11, 2019, approved and formalized the technical standard NTE INEN-ISO 50001 Energy management systems—Requirements with guidance for their use (ISO 50001: 2018, IDT) on a voluntary basis.

Through Official Registry No. 575 of November 11, 2021, Executive Decree No. 229 was published, with which the "General Regulation of the Organic Law on Energy Efficiency" was issued. This regulation aims to develop and structure the necessary regulations to apply the provisions of the LOEE.

Article 18 of the Regulation to the LOEE contemplates that large consumers of energy in commercial, industrial, and public activities will implement the Ecuadorian Energy Management Standard (NTE-INEN-ISO: 50001) in their operations. As of January 30, 2025, they must send their certification to the National System of Energy Efficiency Statistics.[3]

[3]This article was amended in the new Energy Competitiveness Law as follows: Art. 70.- Article 18 is amended to read as follows:
"Article 18.- Implementation of the standard.- Large energy consumers in commercial, industrial, and public activities shall implement the Ecuadorian Energy Management Standard (NTE-INEN-ISO: 50001) in their operations. By January 30, 2025, they must send a certificate to the secretariat of the National Energy Efficiency Committee and subsequently, every three years, they must send an update of the certificate by January 30 of the corresponding year."

6.2.3 Development of the Energy Management Systems Learning Network in the Ecuadorian Industry

This project, promoted by the Ministry of Energy and Mines, is supported by the EUROCLIMA+ Program through the GIZ and aims to promote energy management within the industrial sector of Ecuador as a strategy to improve the energy performance of the participating industries and their contributions to the commitments acquired within the framework of the NDC.

As of 2021, the Ministry of Energy and Mines initiated the approach with companies in the industrial sector to identify the interest in participating in the learning network.

Subsequently, by mid-2022, the Ministry called on industries to form a learning network. As a result of the call, 16 industrial companies with high and medium energy consumption registered for the project. Table 6.1 presents the industrial sectors to which the participating industries belong.

Table 6.1 Branches of the companies participating in the learning network on EMS in the industrial sector of Ecuador.

Sector	Number of companies
Food and beverages	4
Metalworking	3
Plastic	1
Automotive	1
Cement industry	1
Manufacturing—Manufacture of batteries	1
Paper mill	1
Water pumping service	1
Oil and gas	1
Pharmaceutical	1
Textile	1

Source: Ergon Plus, Energía, Agua y Medio Ambiente, S.A. de C.V, Veracruz, México.

The diversity of economic activity of the participating companies has resulted in an exchange of experiences and knowledge that has made the implementation of the Energy Management System more efficient in each company.

The members of the network have managed to form an integrated group based on trust, collaboration, and teamwork, which has allowed fluid communication without limitations in the sharing of experiences. In this way, each company has enriched its implementation of the EMS through the experience of other companies in the project.

6.2.3.1 Methodology of implementation of LEENs in Ecuador

At the beginning of the network's work, coordination meetings were held between the Ministry of Energy and Mines, GIZ, and the implementing company Ergon Plus, where all the documentation generated was shared and reviewed. From that, the concept of implementation, methodology, and schedule was developed, which included the details of all the activities to be developed as part of the process. An analysis of Ecuador's energy regulatory framework and its relations with ISO 500001:2018 was also carried out to identify legal requirements.

The network launch workshop was held in July 2022. Authorities from the Ministry of Energy and Mines, Ministry of the Environment, Water and Ecological Transition, GIZ, the European Union, Ergon Plus, and companies participated. During the workshop, acts of commitment were signed by representatives of industries and ministerial authorities.

Subsequently, a virtual initiation workshop was held, where the members of the network and the consulting team were presented, as well as the scope, objectives, duration, and work schedule. This start-up workshop lasted four hours in a single session.

In addition, the communication channels and digital repository were created, as well as the design of an online monitoring system for the progress of the Learning Network and the implementation of the EMS of each company.

To strengthen the implementing team's action and facilitate iteration with industries, four local technical advisors with experience in energy management were hired. These local technicians were trained in the learning network's implementation methodology, and under the supervision of Mexican experts, they carried out the gap analysis and diagnosis of energy performance for all participating companies.

In all companies, the same methodologies were applied for the survey of energy diagnostics and gap analyses. For this purpose, previous meetings were held between local technicians and technical accompaniers to adjust the methodology and agree on the format of reports for this process.

Local technical advisors were supported by energy management teams or committees from all companies to compile both energy consumption information and equipment inventory. Subsequently, the energy diagnosis report was prepared, following the guidelines of ISO 50002, and the reports were sent to the participating companies in a timely manner to work with them during workshop 3 in the formulation of energy objectives and goals and in the action plans to achieve them.

As part of the learning network process at EMS, eight implementation workshops and four webinars designed to address the specific needs of the network were conducted. The workshops consisted of technical presentations and group activities that allowed them to apply what had been learned and to exchange experiences among the participants.

To define the training topics, a content plan for the workshops and webinars was made based on the diagnoses of training needs and the implementation process of the EMS.

During the workshops, the moderator and the technical companion guided the companies on their way to implement or strengthen their EMS. Work material was delivered, the progress achieved by each company was evaluated, barriers were detected, and actions were implemented to overcome them. The workshops in virtual mode had a duration of 12 hours spread over three days in the same week. Workshop 3 and the closing workshop were face-to-face and lasted two days.

During the operation of the network, there were several opportunities for meetings between the energy management teams of the companies and the technical companions, with the purpose of advising the companies and providing feedback on their work.

It is important to highlight the meetings held with the senior management of the companies, through which their commitment to the implementation of the EMS was strengthened and, in some cases, the interest in obtaining certification in energy management was achieved.

The constant monitoring of the tasks of the companies to advance in the fulfillment of requirements of the energy management system, the elaboration, and implementation of their action plans with specific energy efficiency measures were activities that were carried out throughout the operation of the network, and crucial to achieve the results.

6.2.3.2 Status as of 2023

At the beginning of the project, a gap analysis was carried out in each of the companies' participants, to determine the level of compliance with the requirements of ISO50001:2018. Figure 6.1 shows the results of this activity.

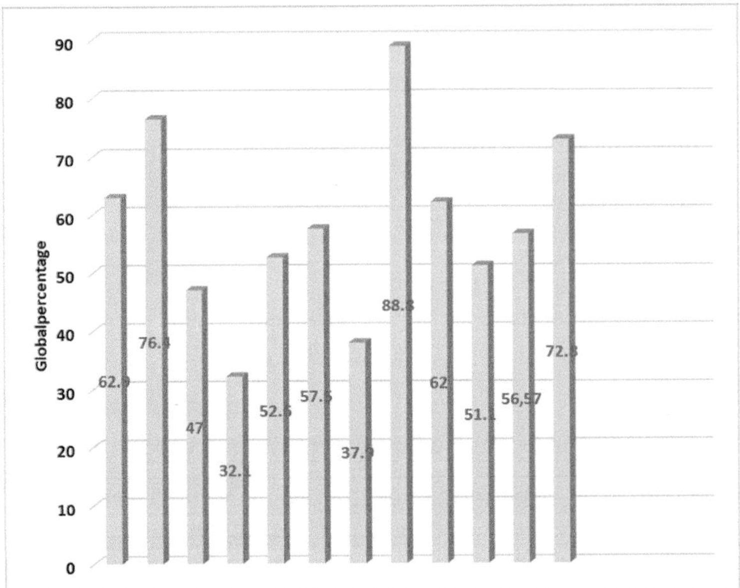

Figure 6.1 Baseline of implementation of the EMS in Ecuador.

Source: Ergon Plus, Energía, Agua y Medio Ambiente, S.A. de C.V, Veracruz, México.

As can be seen, some companies started with an important advance in compliance with the requirements of the standard, while others did so with advances of less than 10%. On average, global compliance with the requirements of the standard of all the participating companies at the beginning of the network was 24.3%.

It should be noted that the company that presented the greatest advance and had already implemented the standard in one of their facilities, so many of the requirements of the standard had already been fulfilled.

The energy baseline was defined as the average energy consumption in the period from September 2020 to October 2022. Table 6.2 shows the energy baseline of each of the participating organizations, as well as that of the set of them. As can be seen, the global energy consumption is 3,029,895 GJ/year, the amount of energy billing amounts to 37,039,760 USD/year, and emissions of GHGs are 201,190 tCO_2/year.

As part of the project, energy performance diagnostics of Level 0 in each of the companies, which allowed them to be characterized energetically, identify their opportunities for improvement, as well as pinpoint the energy benefits, economic and environmental by the implementation of energy efficiency measures. Table 6.3 presents the savings potential identified in the energy diagnosis for each of the participating companies.

The results of the energy diagnostics made it possible to identify 57 measures of thermal energy efficiency and 42 electrical energy efficiency measures, which in total added 99 measures to improve energy performance in the companies of the network, with a total savings potential of 519,034 GJ/year equivalent to 4,908,569 USD/year and a reduction in greenhouse gas emissions by 33,861 tCO_2/year.

6.2.3.3 Learning network implementation as of June 2023

With respect to the percentage of progress in the implementation of the expected EMS to June 2023 by the participating companies, the following are detailed:

- Six companies expect to reach between 75% and 80%
- Five companies expect to reach between 91% and 95%
- Three companies expect to reach between 96% and 100%
- One company did not set a concrete goal for advancement

Table 6.2 Estimated potential electric energy savings of the learning network on EMS of Ecuador evaluated in energy performance diagnostics

No.	Industrial branch	Electricity			Thermal energy			Total		
		Consumption	Billing	Emissions	Consumption	Billing	Emissions	Energía	Billing	Emissions
		kWh/year	USD/year	tCO_2/year	GJ/year	USD/year	tCO_2/year	GJ/year	USD/year	tCO_2/year
1	Balanced feed				56,665	837,596	3,461	56,665	837,596	3,461
2	Automotive	2,915,020	214,907	548	15,442	200,440	943	25,936	415,347	1,491
3	Plastic	9,630,428	621,217	1,810	10,428	170,099	637	45,098	791,317	2,446
4	Aluminum	9,364,442	804,374	1,760	93,691	1,315,177	5,722	127,403	2,119,551	7,482
5	Brewing	14,050,834	913,304	2,640	108,705	792,089	7,968	159,288	1,705,393	10,608
6	Electrical conductors	4,101,048	24,851	771				14,764	24,851	771
7	Coffee mill			-	121,374	1,015,731	8,897	121,374	1,015,731	8,897
8	Textile	2,745,070	205,764	516	19,067	145,087	1,259	28,949	350,851	1,775
9	Processed foods	124,101	15,642	23	5,557	148,350	392	6,004	163,992	415

(*Continued*)

Table 6.2 (*Continued*)

No.	Industrial branch	Electricity			Thermal energy			Total		
		Consumption	Billing	Emissions	Consumption	Billing	Emissions	Energía	Billing	Emissions
		kWh/year	USD/year	tCO_2/year	GJ/year	USD/year	tCO_2/year	GJ/year	USD/year	tCO_2/year
10	Electrical panels	466,475	51,963	88	3,659	72,528	236	5,338	124,491	324
11	Water and sanitation	1,743,677	100,012	328				6,277	100,012	328
12	Oil and gas	8,960,532	931,836	6,476	957,270	8,790,039	70,168	989,528	9,721,875	76,644
13	Paper mill	48,527,801	3,882,224	9,118	908,015	11,515,287	66,558	1,082,716	15,397,511	75,676
14	Automotive batteries	7,959,000	583,126	1,495	178,601	112,705	1,377	207,253	695,831	2,872
15	Cement mill	42,584,640	3,575,413	8,002				153,305	3,575,413	8,002
TOTAL		**153,173,068**	**11,924,633**	**33,573**	**2,478,472**	**25,115,127**	**167,617**	**3,029,895**	**37,039,760**	**201,190**

Source: Ergon Plus, Energía, Agua y Medio Ambiente, S.A. de C.V, Veracruz, México.

Table 6.3 Identified saving potential

Industrial branch	Benefits			Economic evaluation	
	Energy	**Economic**	**Emission reduction**	**Investment**	**Pay-back**
	GJ/year	**USD/year**	**tCO_2/year**	**USD**	**Year**
Automotive	6,743	120,307	454	31,130	0.259
Food	7,447	267,496	901	364,870	1.364
Plastic	3,367	58,476	181	192,043	3.284
Metalworking	846	16,842	47	10,795	0.641
Food and beverage	2,653	32,016	170	59,217	1.850
Metalworking	5,813	121,174	303	559,525	4.618
Food and beverage	8,106	72,653	625	103,520	1.425
Textile	4,362	38,939	266	43,535	1.118
Food and beverage	393	10,473	22	51,573	4.924
Metalworking	824	15,190	77	36,591	2.409
Water and sanitation	2,997	51,949	156	263,754	5.077
Oil and gas	454,599	3,820,467	29,901	4,219,974	1.105
Paper and byproducts	10,286	149,606	331	385,800	2.579
Manufacturing of batteries	4,301	25,544	98	46,560	1.823
Cement mill	6,297	107,438	329	22,500	0.209
Total	**519,034**	**4,908,570**	**33,861**	**6,391,387**	**32.685**

Source: Ergon Plus, Energía, Agua y Medio Ambiente, S.A. de C.V, Veracruz, México.

To achieve 100% implementation of their EMS, six companies set a October 2023 as a deadline: three in December 2023, two in March 2024, and four by June 2024.

Four of the 15 participating companies have established to certify their EMS with ISO 50001: 2018 until June 2023 and the remaining between November 2023 and March 2024.

Based on the identified areas of opportunity for energy performance improvement during the energy diagnostics, each participating organization established its energy objectives and targets and with the set of individual energy objectives, the overall energy objectives of the learning network were established for three stages in time: July 2023 (at the end of the learning network project), to December 2023, and to December 2024.

Table 6.4 presents the overall objectives of the learning network. In all cases, the objectives are both in terms of the corresponding units and in percentage terms with respect to the overall value of the 15 participating organizations.

Table 6.4 The overall objective of the learning network

The project Raw Materials and Climate				
Date	**kWh/year**	**GJ/year**	**tCO_2/year**	**USD/year**
At the end of the LEEN (July 2023)	863,275	17,383	1,610	348,994
	0.56%	0.70%	0.80%	0.94%
December 2023	3,957,089	29,043	3,229	1,017,556
	2.58%	1.17%	1.60%	2.75%
December 2024	6,817,201	88,442	7,086	2,267,931
	4.45%	3.57%	3.52%	6.12%

Source: Ergon Plus, Energía, Agua y Medio Ambiente, S.A. de C.V, Veracruz, México.

As can be seen, the global objectives of the learning network for July 2023 consist of:

- Reduce electricity consumption by 863,275 kWh/year
- Reduce thermal energy consumption by 17,383 GJ/year
- Reduce GHG emissions by 1,610 tCO_2/year and
- Reduce energy expenditure by 348,994 USD/year

It should be noted that the longer-term action plans prepared by the participating organizations will be finalized in December 2024, the date for which the global objective is to be met, with respect to the energy baseline (consumption from the years 2021 to 2022), so:

- Reduce electricity consumption by 6,817,201 kWh/year
- Reduce thermal energy consumption by 88,442 GJ/year
- Reduce GHG emissions by 7,086 tCO_2/year
- Reduce the amount of energy billing by 2,267,931 USD/year

6.2.3.4 Status of implementation of the EMS as of March 2023

During the sixth implementation workshop, held at the end of March 2023, the teams of the participating organizations conducted a self-assessment exercise to determine their degree of progress with respect to the implementation of their EMS in accordance with ISO 50001:2018.

In this exercise, the same gap analysis tool used to determine the baseline was used. The result of this activity is shown in Figure 6.2.

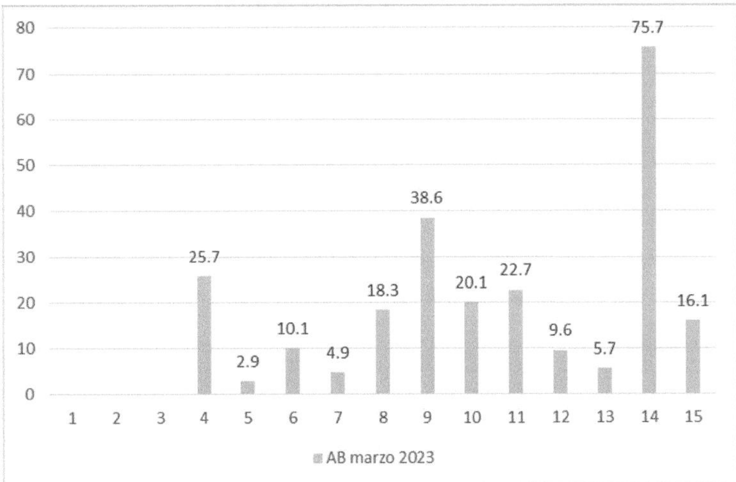

Figure 6.2 Progress in the implementation of the EMS (% vs Companies).

Source: Own elaboration.

Based on the measures to improve energy performance identified in the energy diagnostics, each company, depending on the availability of resources, began with its implementation.

6.2.4 Evaluation of the Experience in Ecuador on Learning Networks in Energy Management Systems

As of March 31, 2023, 60% of the companies in the Network implemented at least one measure to improve energy performance, achieving the implementation of a total of 21 improvement measures, equivalent to 26% of the measures to be carried out by the participating companies.

The other 40% of the companies were also working on the implementation of improvement measures, although until that date they had not concluded with the implementation of any of the measures.

The energy, economic, and environmental benefits of the energy efficiency measures carried out by 60% of the companies in the network are shown in Table 6.5.

As can be seen, the investment made by the companies that implemented the 21 improvement measures, amounted to a total of USD 644,366, and the recovery time of said investment is 1.31 years.

The energy savings amount to 22,689.50 GJ/year and the economic savings achieved correspond to USD 493,339.29 with a reduction in GHG emissions of 2,390.47 tons of CO_2eq.

For each of the companies on the network, individual consultancies were carried out, most of them virtual and, during the missions assigned to the country, face-to-face missions were held, in which the technical companion, the coordinator, and the moderator as appropriate, held meetings to review different aspects of the project.

As of March 31, 2023, 49 individual consultancies had been carried out to ensure that the inputs worked by companies with respect to their EMS, met the requirements of ISO 50001. These consultancies were additional to the work carried out in the workshops and webinars of the project.

Table 6.5 Benefits and economic evaluation of the implementation of measures to improve energy performance

		Benefits			Economic evaluation	
No.	**Industrial branch**	**Energy**	**Economy**	**Emission reduction**	**Investment**	**Pay-back**
		GJ/year	**USD/year**	**tCO_2/year**	**USD**	**Year**
1	Automotive	1,825.10	40,065.28	115.81	9,834.28	0.25
2	Food	1,516.69	161,712.98	636.28	239,320.00	1.48
3	Plastic	834.04	15,059.12	43.53	11,572.00	0.77
4	Metalworking	215.86	4,454.38	13.19	-	-
5	Beverage	5,428.00	68,712.92	370.76	23,016.00	0.33
6	Food	4,505.11	31,042.75	354.00	11,520.00	0.37
7	Textile	447.05	7,512.38	26.13	29,000,00	3.86
8	Food	213.65	4,279.48	11.15	40,103.78	9.37
9	Cement	7,704.00	160,500.00	819.62	280,000.00	1.74
Total savings		**22,689.50**	**493,339.29**	**2,390.47**	644,366.00	**18.17**

Source: Ergon Plus, Energía, Agua y Medio Ambiente, S.A. de C.V, Veracruz, México.

6.2.5 Future Expectations of the Evolution of the LEENs

6.2.5.1 Expectations for improved energy performance

For the achievement of group objectives and goals, the 15 organizations participants in the learning network have proposed to implement a total of 77 measures of improvement of energy performance (21 measures are already in implementation), of which, 31 (40%) will be implemented before the end of the work of the learning network in July 2023, 28 (36.4%) will be implemented by the end of 2023, and the remaining 18 (23.4%) will be fully implemented by the end of 2024.

Of the total of 77 measures, 34 (44%) correspond to thermal energy-saving measures and 43 (56%) correspond to measures to save electricity. Of the thermal energy saving measures, it is planned to implement 13 of them (38%) before completing the learning network project.

Of the measures to save electricity, it is planned to implement 18 of them (42%) before the end of the learning network project. Future expectations for improved energy performance are shown in Table 6.6.

Table 6.6 Future expectations for improved energy performance

	Global goals				Investment	Pay-back
Trimester	kWh/year	GJ/year	tCO_2/year	USD/year	USD	Years
Jan-Mar 2023	36,483	6,891.01	774.01	164,416	255,497.60	1.55
Apr-Jun 2023	863,275	17,383.08	1,609.50	348,994	385,475.77	1.10
Jul-Sep 2023	2,808,436	20,114.32	2,491.62	516,676	810,667.09	1.57
Oct-Dec 2023	3,957,089	29,043.44	3,229.05	1,017,556	1,413,197.00	1.39
Jan-Mar 2024	5,309,935	41,491.92	4,147.57	1,213,567	1,794,124.12	1.48
May-Jun 2024	6,458,034	88,154.22	7,626.30	2,016,153	3,286,312.82	1.63

	Global goals				Investment	Pay-back
Trimester	kWh/ year	GJ/year	tCO_2/ year	USD/ year	USD	Years
Jul-Sep 2024	6,812,480	88,154.22	7,696.44	2,042,265	3,508,854.11	1.72
Oct-Dec 2024	6,817,201	88,441.86	7,086.32	2,267,931	5,911,125.21	2.61

Source: Ergon Plus, Energía, Agua y Medio Ambiente, S.A. de C.V, Veracruz, México.

6.2.5.2 Expectations of ISO 50001:2018 certification

All participating companies have expressed their interest in obtaining the ISO50001:2018 certification. Four of them have already started negotiations to be certified until June 2023 and the rest have expressed their intention to do so between November 2023 and March 2024.

During the seventh implementation workshop held in April 2023, the 14 participating companies said they would like the network to continue and agreed to seek mechanisms to make it possible.

6.3 The Cuban Experience in Learning Networks

Angel Alfredo Núñez Barreto[a] and Welner Collejo Jerez[b]
[a]*Director, National Office for the Control of the Rational Use of Energy (ONURE), Matanzas, Cuba*
mt.lopez@nauta.cu, dtorure@electmtz.une.cu
[b]*Deputy Director, National Office for the Control of the Rational Use of Energy (ONURE), Matanzas, Cuba*
welnerc@oc.une.cu, wcollejo1980@gmail.com

In October 2018, as part of the Triangular Cooperation between Germany, Mexico, and Cuba, to strengthen institutional capacities for the implementation of actions of the policy of renewable energy sources and energy efficiency in Cuba, a workshop was held in Havana for the planning of a program for Energy Management

Systems with the participation of specialists from the National Office for the Control of the Rational Use of Energy (ONURE) and National Organizations, being taught by experts from the National Commission for the Efficient Use of Energy (CONUEE) of Mexico.

The workshop allows learning, from the Mexican experience, a methodology of capacity development and interaction of people and institutions seeking to implement energy management systems and energy efficiency measures. To achieve this, the exchange of experiences and the technical support provided by experts in the field is called Learning Network.

In the workshop, the group identified several learning networks, which in some cases would be formed by companies from different sectors and in others by companies or institutions in the same sector. It was when the idea of making a learning network with high energy-consuming hotels in the Varadero Tourist Pole arose, which in time became *the first Cuban experience.*

Between November and December 2018, directors of ONURE in the province of Matanzas and the Delegation of the Ministry of Tourism in Varadero exchanged, analyzed, and agreed on the development of the Learning Network preceded by the self-preparation of ONURE Matanzas specialists, based on the experience of CONUEE. It was then that the kick-off workshop of the LEEN took place on April 29, 2019, in the Plaza America Convention Center, where ONURE, as the initiating organization, oversaw the welcoming words to the event.

After the presentation of the different actors of the network, the Collaboration Agreement between the Ministry of Tourism in Matanzas (MINTUR) and ONURE was signed; an important document that defined the roles and responsibilities of each of those involved. Previously, the Collaboration Agreement was read and the representatives of each of the actors involved proceeded to sign it.

The background and criteria that were considered for the planning of the network with the MINTUR in Matanzas were that this sector has an important weight in the energy matrix of the province, the awareness of its staff on energy and environmental issues, the stability and professional level of the personnel in charge of managing energy, and the geographical proximity of the potential participants.

The first of the seven workshops held took place on May 29, 2019. The results of the Initial Energy Diagnostics in 10 organizations are shown in Table 6.7.

Table 6.7 Quantitative results of the initial energy diagnostics

Average annual consumption MWh	Annual savings potential MWh	Organizational measures to be implemented	Maintenance measures to be implemented	Investment measures to be implemented
73152	5148	8	5	48

Source: ONURE.

The expenses in energy carriers and water represented percentages ranging from 9% to 15% of the total expenses of the organizations. The air conditioning, impulsion, distribution, storage of fluids, and the production of domestic hot water were the systems with the highest energy consumption. In this first workshop, the goals were established, for which three organizations committed to certify their energy management system and the remaining seven to implement, showing the commitments.

Between May 2019 and February 2020, six workshops were developed. Then the paralysis of the learning networks due to COVID-19 appeared, resuming the seventh workshop in November 2020.

The work of the workshops consisted of the exchange and orientations aimed at the organizations to define the best possible way the context, the scope of their management system, their energy policy, their goals, the baseline, their energy performance indicators, and the other aspects of numeral 4 of the Cuban Standard ISO 50001.

Technical issues related to efficiency in the energy systems of organizations (climate systems, fluids, etc.) were studied and analyzed in the workshops, including the benefits of renewable energy sources between workshops, support, and evaluation visits were carried out on the tasks indicated in the workshops, the energy performance of the main indicators, and the savings opportunities identified in the energy reviews by the initiators and technical experts.

The third workshop held in July 2019 presented the new Cuban Standard ISO 50001: 2019 and the main changes with respect to 2011.

The closure of this network occurred on February 8, 2022, and at that time 8 of the 10 participating organizations were considered with their management system implemented. The most repeated criteria in the closing plenary include an unprecedented socialization that allowed unity of criteria and exchange of experiences. It was achieved that the energy workers worked with management systems and that the workshops and activities developed had a lot of professionalism, dynamism, and high technical level, and doubts were clarified, achieving a successful and happy term.

The Learning Network had 44 participants, including 17 women (38%), in addition to four technical experts from the University of Matanzas, five sponsors of the Tourism Delegation, and six specialists and managers of ONURE who assumed the roles of moderators and experts in addition to being the initiators of the network.

Among the impacts achieved were the replacement of 98% of both internal and external lighting to LED with a saving of 54,500 kWh and the non-emission of 45 tons of CO_2, the acquisition of refrigeration and air conditioning equipment with Inverter technology allowed to reduce consumption by 3,222 kWh with 3 tons of CO_2, and the sealing of 100% of the cold rooms meant 9,530 kWh and 8 tons of CO_2 not emitted within a year at the Meliá Varadero Hotel.

The Sol Palmeras Hotel among its main results include the replacement of 50 electric heaters with solar support representing a saving of 5 kWh per day per equipment, which per year 91 thousand 250 kWh and 76 tons of CO_2 stopped emitting and the acquisition of a more efficient dishwasher decreases daily consumption by approximately 37 kWh per day that per year are 13 thousand 505 kWh and a reduction of 11 tons of CO_2.

The establishment of the Energy Policies of the organizations had its impact, among which the Iberostar Selection Varadero Hotel stands out. It is based on the commitment of its management to ensure total customer satisfaction, exceeding their expectations, and leaving them unforgettable memories. Complying with the demands and standards of the Iberostar Chain, based on compliance with all laws and regulations of our country, and continuously improving the energy performance of the organization. We comply with our policy, focused on the continuous improvement of processes.

Counting on the active participation and professional, historical, and cultural training of our workers; achieving to be included among the leading hotels in Cuba. The purpose of being the leading hotel of customer satisfaction in Cuba, as well as directing its actions to meet the requirements of our relevant stakeholders. Therefore, senior management provides all human and material resources and undertakes to implement, document, maintain, and continuously improve an Integrated Quality and Energy Management System, by the NC ISO 9,001: 2015 and NC ISO 50001: 2018 standards and with the following objectives:

- Comply with the legal and regulatory requirements applicable to the organization
- Periodically verify compliance with its quality and energy objectives, taking the relevant actions based on the results found
- Ensure the acquisition of products and services with maximum energy efficiency and friendly to the environment
- Promote the use of renewable sources of energy
- We act so that our processes are increasingly cleaner and to be able to provide products and services to our customers that increase their level of satisfaction
- Regularly update our policy and communicate it to all staff for their conscious participation in improving energy performance

The city of Cárdenas in Matanzas where this first Learning Network was developed is known as the city of firsts, among other things, because it was in it where for the first time in Cuba, the flag of the lone star waved, the first public lighting company in Cuba was established in it. The first oil well-stranded at sea was found, being then this first Cuban experience of a Learning Network for the Implementation of Energy Management Systems was a new first for Cárdenas.

This first experience of the Tourism Network had its replica in the province of Matanzas itself with the participation of other organizations as detailed in Table 6.8.

Table 6.8 Detail of participants in the province of Matanzas

Network name	Participating organizations	Municipality	Persons involved	Women	State
Industry	2	Matanzas Cárdenas	33	4	Ended
Gaviota Hotel Chain	14	Cárdenas	70	25	Ongoing
Food industry	2	Cárdenas	18	4	Ongoing
Manufacturing and food industry	2	Jovellanos	18	5	Ongoing
Total	**20**		**139**	**38**	

Source: ONURE.

The networks have a presence in other provinces of the country, as shown in Table 6.9.

Table 6.9 Detail of participants in other provinces of the country

Network name	Participating organizations	Municipality	Persons involved	Women	State
Tourism	4	Pinar del Rio	7	2	Ongoing
Food industry	2	Pinar del Rio	14	3	Ongoing
Food industry	4	La Habana	21	7	Ongoing
Mix	8	Cienfuegos	31	8	Ongoing
Industry	7	Villa Clara	31	14	Ended
Food industry	6	Villa Clara	31	7	Ended
Mix	5	Ciego de Ávila	15	6	Ongoing
Tourism	8	Granma	11	0	Ongoing
Total	**44**		**161**	**47**	

Source: ONURE.

It has been considered that 17 organizations have their management system implemented. In the process of certification, three certified with the National Office of Standardization in Cuba that are: Hotel Blau Varadero in Matanzas, the Central Rum Factory in Villa Clara, and the Cienfuegos Generation Company, being the first two candidates to certify with AENOR Mexico.

What has been achieved so far has been possible due to the experience of CONUEE, the delivery of ONURE specialists, and the dedication of the organizations involved, which is a challenge to continue multiplying in favor of the integration of energy efficiency, management systems, and the necessary environmental protection.

6.4 The Colombian Case[4]

The Colombian Energy Efficiency Learning Networks, launched in 2022, are part of a strategy by the Ministry of Commerce, Industry, and Tourism (MinCIT), together with the Ministry of Environment and Sustainable Development (MinAmbiente) and with the support of the European Union and the Federal Ministry for Economic Cooperation and Development (BMZ) through the Euroclima Program. The initiative is implemented by GIZ and local implementers.

According to the government, this initiative is an important component of strategies to improve the sector's energy performance and competitiveness. It also contributes to the mitigation targets adopted in Colombia's Nationally Determined Contributions (NDCs). They also aim to promote business sustainability and contribute to the national objectives of the Strategic Line on Energy Efficiency of the Comprehensive Management of Climate Change.

6.4.1 The Context of the Implementation of the Colombian LEEN

The Euroclima programme is funded by the European Union and co-financed by the German government through the Federal Ministry

[4]This section is based on information provided by the Deutsche Gesellschaft für Internationale Zusammenarbeit (GIZ) GmbH, especially on the Summary of the Final Report for the Project.

for Economic Cooperation and Development (BMZ). This program is linked to the partner countries through the Country Dialog process, in which climate change and biodiversity priorities are identified, while actions to be implemented are defined.

In September 2020, Colombia submitted its NDC to the UNFCCC, committing the country to a 51% reduction in greenhouse gas emissions by 2030, compared to a business-as-usual scenario.

In this context, in 2020, Minambiente requested the start of the Country Dialog process, which resulted in the prioritization of two specific actions in the industrial sector (one of the most relevant sectors in terms of energy demand and GHG emissions). Therefore, within the framework of Action 8: *"Energy management in the industrial sector as a strategy to improve the energy performance and competitiveness of the sector and contribute to the mitigation objectives adopted in the NDCs,"* the process for the creation of the country's first LEEN was initiated.

The main objective of the creation of this first network was to have a pilot experience that could be replicated at the national level, as well as transfer the methodology of the LEEN to domestic stakeholders and the adoption of this strategy.

Initially, the LEEN included 16 big and medium leading companies from different sectors of the Colombian economy. Finally, 13 companies participated in the whole process: Cartón de Colombia, Compañía Nacional de Levaduras, Industria Nacional de Gaseosas, Compañía Nacional de Chocolates, Bio D, Sociedad Industria de GrasasVegetales, Empacor S.A, Esenttia S.A, Grupo Siderúrgico Reyna SAS, Nuvant S.A, Colombina del Cauca, and Holcim Colombia S.A y Ladrillera Helios S.A.

6.4.2 Methodology of Implementation of the LEEN in Colombia

The guiding conceptual framework for this LEEN was the methodology proposed in the guidebook *"Guide for the Implementation of Energy Efficiency Learning Networks and Energy Management Systems in the Latin American Context,"* published by the Deutsche Gesellschaft für Internationale Zusammenarbeit (GIZ) GmbH in September 2016. Nevertheless, some adaptations to the local context and modifications were performed, which were important in terms of the impact on the performance of the LEEN.

The first Exchange Workshop and a Start-up Workshop were held, followed by five Implementation Workshops and, finally, a Results Consolidation Workshop and Protocol Ceremony.

As a complement to the workshops, four Open Spaces were implemented, in which technical topics, prioritized by the network members, were discussed in depth. Prioritization was initiated in the start-up workshop and was completed in the First Implementation Workshop. Some prioritized topics were efficiency in steam systems, measurement, and performance indicators, EERR-sustainable technologies, fuel management and efficiency, and efficiency in cooling systems.

Regarding the team and professionals implementing the LEEN, it is important to highlight that it was the result of the integration between Colombian and other Latin American experts and consultants. Indeed, the participation of local organizations, in the elaboration of energy diagnoses and gap analysis, has been a key characteristic of this case. The local organizations participating were Azimut Energías S.A.S., the Colombian Energy Efficiency Council (CCEE), and the Corporación Ambiental Empresarial (CAEM), which help with the fluid relationship.

As described in the Summary of the Final Report (GIZ, 2022), this LEEN used participatory and systemic strategic planning methodologies, based on a combination and adaptation of situational strategic planning methodologies[5] (ZOPP is the acronym for the German expression "Zielorientierte Projektplanung" or objectives oriented project planning) (Helming and Göbel, 1998) and change cycle[6], complemented with the application of tools for complex systems in situations of high uncertainty with multi-stakeholder referents. In the participative workshops, other tools used were strategic management, administration, planning, process and group evaluation, and project cycle tools used in combination with the METAPLAN visualization technique.

[5]http://www.cepal.org/ilpes/noticias/paginas/2/36342/LIBRO_ENTREVISTA_CON_MATUS.pdf.

[6]Anke Kaulard/ Karin Neck, serie investigaciones 2. La gestión de cambio para una mayor eficiencia de la promoción del Desarrollo económico local desde los gobiernos locales y regionales, accedido el 17 de septiembre de 2015 por http://kauco.de/wp-content/uploads/Gesti%C3%B3n-del-cambio-y-DET.pdf. Adaptación del Modelo de gestión de Cambio Organizacional, Denkmodell.

6.4.3 Results Arising from the Colombian LEEN[7]

As a group commitment, companies plan to make overall energy savings of between 2% and 7%, which is significant for these highly energy-consuming companies; nearly 267,968 MWh/year, which translated into GHG emission reductions (according to the official Colombian emission factors) will result in 50,595 tCO_2/year[8]. As highlighted by the GIZ expert, Yaklan Andrea Zapata Gomez, this commitment is very relevant because the goal of energy efficiency and energy management in Colombia is 1.7 $MtCO_2$ eq./year.

Table 6.10 shows the targets disaggregated by source. It is important to highlight that by the end of 2023, the implementation of EMS was close to 85% and two companies are expected to certify.

Table 6.10 Global commitments for the EMS LEEN in Colombia

Fuel	Energy consumption (GWh/year)	Saving (%)	Saving (GWh/ year)
Power	8,448	2.1	173.2
Coal	1,603	2.1	82.6
Natural gas	495	5	10.5
Others	93	1.8	1.7
Total	**10,639**		**268**

Source: Adapted from GIZ (2022).

Moreover, as stated in the final report of the LEENs, the results have been very satisfactory. All the 13 companies participating have completed the process of the LEEN and they have achieved interesting progress in terms of implementing opportunities for improvement, energy savings, and improved energy performance, as well as significant progress in terms of implementing EMS. They have also expressed their interest in continuing to work together with the rest of the companies that make up the LEEN.

An additional benefit highlighted by the implementers is the potential multiplier effect of the methodology, as many of the

[7]Some important information on results from the Colombian LEENs can be seen in this video: https://www.youtube.com/watch?v=3dxGVV3RfQU.

[8]See: https://www.euroclima.org/actualidad-ec/noticias-es/2013-empresas-de-la-red-de-aprendizaje-de-eficiencia-energetica-se-comprometen-con-la-disminucion-de-emisiones.

companies that participated in the process have other plants in Colombia, to which they could transfer the knowledge acquired throughout the process.

6.4.4 Lessons Learnt from the Colombian Case

As stated in the Final Report of the Colombian LEEN, a lot of lessons arise from the first experience in the country.

The first aspect to highlight refers to the workshops, an essential part of this methodology as seen in Chapter 2. In this case, LEEN has followed an integrated methodology, virtual, and on-site. However, according to the implementers of the LEEN, at least three of the workshops should be developed in person: the initial workshop, an intermediate workshop, and a results presentation workshop. The initial workshop is key because it is the moment at which the participants get to know each other and strengthen ties, which is fundamental for the future performance of the network. If trust is not achieved, participants will be willing to share experiences and information. They also learn about the methodology and some technical aspects of ISO 50001, etc. during the first workshops, which increases the relevance of an in-person experience. It is also important to highlight the role of methodologies and tools used for communication (WhatsApp group, individual e-mails, invitations via Teams, and community Drive), which are key to maintaining regular contact between participants.

Then, the roles of each actor within the network must be well structured. In this regard, the coordinator is responsible for maintaining the link between the various actors in the LEEN, overseeing the proper integration and transfer of information, and promoting the proper management of the activities involved in the implementation process. The moderator must promote communication and motivation among the members through the proper implementation of work methodologies and dynamics. Technical experts are the one who must guide and assist the members of the network in the proper implementation of the EMS, in compliance with the requirements of ISO 50001, as well as the constant search for energy performance improvement, through technological measures, and good operation and maintenance practices. However, because of this first experience, the implementers highlight that it may be important to incorporate interns (advanced engineering students) to assist with the tasks of collecting and organizing information, data collection, document organization, etc.

In this case, the materials and forms used to carry out the Energy Performance Diagnosis and Gap Analysis were developed jointly by the implementation teams of the LEEN from Colombia and Ecuador, and they were very important because allowed a more accurate and equitable guidance between the different local experts.

It is advisable for future experiences to enhance the idea of the relevance of the information and documentation that companies should make available for Energy Performance Diagnosis and Gap Analysis. It is also recommended to provide specific spaces to support the companies in this task, to avoid delays and achieve better results.

In the case of Colombia, the setting of individual and collective goals was hindered by the fact that some of the reports were not available in due time and form, in addition to a certain disparity in the results, depending on the different local advisors. In this regard, it is suggested that the Technical Expert guides the development of the Gap Analysis, which should be done by the participants (they are the ones who know their organizations). In addition, it is suggested to start Energy Performance Diagnosis early.

A novelty of the Colombian LEEN that was very successful was the Open Spaces, spaces for specific topics defined and prioritized by the members of the network, in which the participation of colleagues from other countries was seen as very positive.

Another key aspect of this experience was the meetings with the senior management of each of the companies, which have contributed to the empowerment of the LEEN and the visibility of its work and results while facilitating the availability of resources (human, technological, economic, etc.) to achieve improvements in terms of energy management and energy performance.

Finally, an enriched experience has resulted from the joint work between colleagues of the Ecuadorian and Colombian LEENs, both from a technical point of view and a human one, since it has favored the exchange, the unification of criteria, and the availability of resources and methodologies for moderation, technical support, and coordination of both networks.

6.5 The Nicaraguan Case

Claudio Carpio
Based on an interview with Jairo Fernando Calvo, a consultant on energy efficiency and an expert in the Nicaraguan case
seifercalvo@gmail.com

To include a variety of LAC experiences, a semi-structured interview was conducted with Jairo Fernando Calvo, one of the technicians involved in the LEEN experience in Nicaragua, to obtain some of the main lessons learned from this case.

The interview is transcribed as follows:

1. **What energy efficiency policies (and/or programs) have been carried out in your country?**
   Nicaragua has an Energy Efficiency Law, Law# 956, approved in June 2017. Published in "La Gaceta," Official Gazette # 128. Literally Article 3 Competence and Responsibility of the State, says: *"The State, through the Ministry of Energy and Mines (MEM), has the responsibility to organize, formulate, promote and ensure the development and monitoring of national programs for the rational and efficient use of energy in accordance with the provisions of this Law. The State shall allocate the necessary resources through competent bodies for the development of these plans and programs."*
   To date, little or nothing of this Law is reflected in the Government's actions toward all public/private entities that are part of the engine of the Nicaraguan economy. Energy efficiency is not considered important despite the results obtained in the learning network.
2. **What productive sectors are addressed or have been addressed?**
   The law in its Article 9, "Sectoral working groups," says: *"The MEM will be able to set up sectoral working groups to share information and coordinate strategies with the different actors to promote the rational and efficient use of energy in the commercial, industrial, transport, government, and education sectors, among others. The coordination of all the working groups will oversee the MEM and participation in them will*

be mandatory for the public sector, and voluntarily with the private sector interested in participating." This law has not had effect, Article 9 is no exception. Most of the efforts made to date have been made by non-governmental entities with the support of foreign cooperation.

3. **What have been the main barriers or problems of energy efficiency in these sectors?**

   Because companies have innumerable challenges to produce and add to the Nicaraguan context, they are focused on producing to the maximum and solving the most pressing problems, forgetting, or taking a back seat to energy costs.

4. **What has been the context of the emergence of learning networks?**

   The appearance of the ISO 50001 standard, although it should be the easiest to adopt by most companies, due to its quick way of being able to make the efforts tangible, is not because at first it was considered one more standard that comes to swell a large list of standards related to the industrial sector.

   Learning networks were crucial to making the implementation of ISO 50001 faster, more technical, and more objective, helping plant managers to find savings and reduction in the implementation times of new technologies, among many other things.

5. **How was the arrival of the learning networks in your country?**

   The Chamber of Industries of Nicaragua (CADIN) had already been working closely with German cooperation on many issues in the field of energy efficiency. It was German cooperation, through a triangular cooperation project in which Germany was involved through (GIZ), Mexico through the Ministry of Energy and the National Commission for the Efficient Use of Energy of Mexico, (CONUEE), El Salvador through the Salvadoran Association of Industrialists (ASI), and Nicaragua through the Chamber of Industries of Nicaragua, who participated in the project *"Implementation of an Energy Management System Based on ISO 50001"* through learning networks.

6. **What is the role of cooperation between countries in bringing the experience from one country to another?**

   In the case of Nicaragua, without the technical-economic support of foreign cooperation, this experience would take us many years to acquire.

7. **What is the role of the specific institutional structure in the performance of the networks?**

   Although each country has its characteristics and learning networks have something in common with all countries, they need entities and people who enjoy prestige and credibility when carrying out the initiatives so that they meet the interests and objectives projected.

   Thus, the role is to give correct control and monitoring of all the activities that result in the interests and objectives of the project.

8. **What is the role of the main companies or the main consumers in these networks?**

   The companies participating in the learning networks must be aware that improvements take time and, in turn, must guarantee the necessary time and resources to the participants so that they can achieve the objectives set at the beginning and those that arise with the course of the network.

9. **How did the international experience of the networks adapt to the specific case in your country?**

   Because it is the first network where many professionals share information and experience without reluctance, it takes time to understand that acquired knowledge not shared is not knowledge (Fernando Calvo). Well, it only serves one, and when it is no longer there.

   It was not necessary to adapt, but rather to make known information and experience that led them to the knowledge of actions and initiatives that could be done faster and better with the help of another who had already experienced.

10. **How was the reaction of the private sector? Was participation achieved?**

    Any change in the way of doing things generates distrust, at first, and risks without knowing the results are few. The choice of these companies was in charge of CADIN and was made with companies with which it was already working and that had confidence in those who proposed participation in the network.

11. **What results were obtained with the implementation of the learning networks?**

    More than the excellent economic results obtained through the network, it could be seen that it has multiple applications in all fields of daily life and can help shorten the time it would take to make improvements in the usual way.

12. **What are the main challenges identified when implementing the networks and how they were overcome?**

    Sharing information from companies about how it is being produced with unknown people everywhere represents a challenge, especially since the thought is that we will give valuable information to the competition.

13. **Were there positive externalities in this process, e.g., gender component and/or participation of any academic institution?**

    The gender component for us is basic and is immersed in all our projects.

    As soon as the project was finished and the results were announced, many educational institutions and NGOs approached us, to see what we could do including the government. But due to the events in the country since April 2018, everything took a back seat and lost all importance.

14. **What would you propose to continue the implementation of learning networks in your country?**

    There must be a real commitment on the part of the government to the implementation of learning networks in all areas. Such a tool will bring many tangible and non-tangible benefits to the economy. Cooperation and companies have already done their part, the results are there in sight and only continuity should be given.

15. **Final comments, etc.**

    Learning networks help us understand that in the world we are all connected and trying to solve problems... what we do not know is that others have already done it or at least have tried and that it is easier to continue what they have done than to start from scratch. However, we do not realize countless factors, forgetting that we are all connected and that several heads think better than one.

Learning networks have broken down many barriers so that we can share experiences and knowledge without fear. Emphasis is placed on this because many confidentiality agreements had to be signed for fear of companies.

Finally, just mention and thank all the people and companies that participated in the first learning network in Nicaragua so that they could see that learning networks do leave a productive legacy for them, for companies, and for institutions. And that if at any time a group of professionals, companies, or government entities decide to participate in a new learning network, they will be proof that they will not waste their time.

Chapter 7

Final Reflections

Marina Yesica Recalde[a,b] and Claudio Carpio[a]
[a]*Fundación Bariloche, Argentina*
[b]*Consejo Nacional de Investigaciones Científicas y Técnicas (CONICET), Argentina*
mrecalde@fundacionbariloche.org.ar, claucarpio@gmail.com

In a world seeking mid-term decarbonization pathways, effective and efficient alternatives to reduce the environmental impact of energy consumption have become crucial, and energy efficiency emerges as an important alternative. The empirical evidence has shown that supply-side initiatives will not be enough, per se, to reduce the environmental impact of the energy sector and that the energy efficiency options (either technological change or rational use of energy) are cost-effective. In the specific case of industrial sector, the IEA argues that meeting SDGs will demand industrial emissions fall by 1.2% annually till 2030, demanding for and intensification of efforts in energy efficiency.

Learning Energy Efficiency Networks: Applications and Challenges in Latin American Countries
Edited by Claudio Carpio and Marina Yesica Recalde
Copyright © 2026 Jenny Stanford Publishing Pte. Ltd.
ISBN 978-981-4968-82-9 (Hardcover), 978-1-003-77468-6 (eBook)
www.jennystanford.com

DOI: 10.1201/9781003774686-8

Despite its recognized relevance, improving energy efficiency in industrial sectors of developing countries has been difficult. Firms face significant problems that prevent them from implementing cost-effective actions. Therefore, the implementation of policies, actions, and instruments is required. There are many different alternatives that can be implemented to address these problems, and they depend on the type of barrier to be removed, and on the specific boundary and enabling condition.

In the late 1980s, a novel instrument to remove knowledge and investment barriers in the industrial sector was developed: **Learning Energy Efficiency Networks**. LEENs have shown to be an effective instrument to promote energy efficiency, generate different benefits in the countries in which they are being implemented, and in the companies that have participated in the process. Cooperation among network increases the company's commitment to address goals and targets, and the horizontal flow of information is key to the performance.

The main objective of this book was to discuss some Latin American experiences to identify some similarities and challenges that could be relevant for future extensions. As follows a brief resume will be presented for some of the cases.

7.1 The LA Experiences

Most of the LAC LEENs' experiences have been characterized by a relevant role of German cooperation through the Deutsche Gesellschaft für Internationale Zusammenarbeit (GIZ). Germany, which has been the leader in the implementation of these networks worldwide, has had a key role in this region throughout its Cooperation Agency, both as a financing agency and as a technical capacity builder. Another key aspect has been the intraregional cooperation. For instance, between 2016 and 2018, a cooperation project was implemented between Mexico, Germany (as southern and northern provider partners respectively) and El Salvador and Nicaragua, as the beneficiary partners. As part of this process, two LEENs were successfully developed, one in El Salvador and the other in Nicaragua, which constituted the first experiences for networks in Central America, resulting in significant annual economic savings, and capacity building in both countries throughout courses, and material on energy efficiency.

Mexico is one of the leaders of the region, with 21 LEENs with the participation of 167 public and private companies and/or entities with 254 facilities; almost 1,000 professionals directly trained in topics related to the process of the EMS, and more than 4,000 indirectly; and the participation of a lot of public and/or private organizations related to the efficient use of energy and/or processes related to EMS: ANCE, CAMEXA, AMEXGEN, AMENEER, Universities, Institutes, etc. The Mexican case is characterized by a very flexible strategy and CONUEE's leadership, including different sectors and areas, such as public (federal and municipal) and private (both large companies and SMEs), as well as various end-use sectors of energy (industry, buildings, transport, and municipal services). The documentation of all the activities (videos, manuals, and tools) was useful to replicate the actions within and outside the country. LEENs have been a very appropriate and successful mechanism to introduce organizations to EMS and its application through ISO 50001 in the country. Indeed, in Mexico, the promotion of EMS was based, mainly, on the promotion, creation, and development of LEENs.

Two elements of the LEENs experience and methodology were very important. Firstly, visits to the plants: participants mentioned that these visits allowed them to learn about new processes, from the experiences of other companies, and about new focused activities for the optimization of energy consumption. Secondly, exchanging experiences: as it helps to reduce EMS implementation times, learn from the experiences of other companies, share knowledge, and contact people who are experts in different topics and are willing to help each other.

As discussed by the authors, the Learning Networks methodology, originally imported from Germany by GIZ and incorporated by CONUEE, was well adapted to the Mexican context, and proved to be a powerful tool for capacity building, for the implementation of EMS, and a particularly novel training method for Mexico. Beyond its initiatives and experiences, Mexico has assumed strong regional leadership in this area. In fact, it has technically prepared professionals from El Salvador, Nicaragua, and Argentina.

In 2018, following the Mexican experience and impulsed by the CONUEE, GIZ, and BMZ, Argentina initiated the first LEEN, which was followed by other eight experiences, financed by international organizations and the local government. In this case, networks were born in the industrial sector (including big and small companies), followed by commercial and service sectors.

One important barrier faced in the implementation of LEENs in Argentina was the lack of energy information and the energy culture of organizations which is not prompt to energy efficiency. This is coherent with empirical studies pointing out that information and cultural awareness are some of the most important barriers to energy efficiency in industrial sectors. However, these barriers were surpassed by the LEENs (as declared by most of the participants).

Similarly to Mexicans, participants highlighted the positive impact of workshops as collaborative learning techniques. The workshops provided excellent space for the exchange of knowledge, strengthening capacities of members, generating trust, and cohesion.

Between 2017 and 2021, Brazil developed seven LEENs in the water supply and sanitation services sector, with international cooperation and the support of GIZ. The sector selected for the LEENs made the Brazilian a particularly interesting case as water and sanitation sector are a natural monopoly, without competition, which erases the incentives to implement energy efficiency actions to improve competitiveness.

These LEENs have helped the country to overcome some of the most common barriers to energy efficiency. While knowledge and management barriers were firstly removed in the first 6 LEENs; latter reductions in energy subsidies provided enabling conditions to enhance the relevance of energy efficiency.

One aspect to be highlighted is the criteria for the selection of companies. Organizations were mainly selected by call and preferences were given to the most engaged ones. Then, a midterm evaluation report was issued by the network coordinator, pointing out which tasks were concluded, and which needed to be concluded by the participating beneficiary organization. The evaluation report was sent to the participant's supervisor for information. With the midterm evaluation, the participants were reminded that this learning network was not passive training. Also, a final evaluation was issued.

No admission fees were collected. Costs of were covered by the German Brazilian Cooperation and the local anchor entity. Beneficiary organizations provided venue facilities hosting the meetings and provided input time for their staff. To increase the company's commitment, at the closing session of the events, participants handed out their finished products and outputs, the Water and Energy Management Plans, to their supervisors or even the city mayor. In this way, the mayor also acknowledged the commitment of the team, increased his support for the proposals, and identified efficiency projects in the plan.

In the Chilean Case, the Learning Network of the 4e Renewable Energy and Energy Efficiency program, developed by GIZ to improve energy performance and reduce emissions for mining, were a great challenge, because of the COVID context. The methodology of the LEENs had to be adapted to the existing mining and national reality in such a way that this network could be initiated and achieve the expected results.

Companies have found the information shared, analyzed, synthesized, and delivered to be very valuable and believe that the network has allowed them to work on their energy management system and improve their processes. This helped build trust and confidence in the network, which is of great importance in promoting exchange of information and experiences. In the first year of the network, 14 companies initially participated, in the second one two additional companies joined. Many companies expressed interest in continuing to participate actively in the LEEN. On average, a total of 55% participation was obtained from large-scale mining companies, 40% from medium-sized mining companies, and an average of 2% from small-scale mining companies. Workshops were one of the main challenges. More effort is required to get more companies to participate and want to share knowledge with other companies. Other barriers were the reluctance of the sector and professionals, and the willingness to share knowledge.

This has been the first time worldwide of the methodology implemented in the mining industry, and it managed to achieve the objective set by GIZ and generate a great impact within the mining ecosystem.

Framed in the Diálogo País project, and financed by EUROCLIMA, in 2022, Colombia initiated its first experience in LEENs. It was made up of 13 companies, most of which are energy-intensive companies. It was a very satisfactory experience, as companies are expected to make overall energy savings of between 2% and 7%, which may represent nearly 267,968 MWh/year, and which may result in 50,595 tCO_2/year of GHG savings, a very important target, for a unique activity, considering that the goal of all energy efficiency and energy management actions in Colombia is 1.7 $MtCO_2$ eq/year.

The Colombian experience has been very novel in the fact that it was implemented by an Argentinean consulting group and is extremely connected to the Ecuadorian LEEN. This aspect clearly shows the positive impact of intercountry networks in the implementation of this methodology.

Two additional aspects can be highlighted. First, the role of local institutions as part of the implementers has enhanced the applicability of the instrument. Second, the Open Spaces activities have been created to give responses to some topics that the participants highlighted to be important. Furthermore, the lessons learned from this first experience in the country are very well documented in the final report of the project and could be considered in detail in the future applications and implementations of the LEEN in the rest of the country, as mentioned by the consultants. This is of strong relevance, as the document mentions that the LEENs could be deployed in different regions of the country.

Cuba has experienced the issue of networks in the tourism sector, through the Ministry of Tourism, as part of the Triangular Cooperation with Germany and Mexico.

The main criteria selection in this case was the weight in the energy matrix of the province, the awareness of its staff on energy and environmental issues, the stability and professional level of the personnel and the geographical proximity of participants. At the end of the network, February 2022, 8 of the 10 participating organizations were considered to have their management system implemented; 98% of both internal and external lighting was replaced by LED, refrigeration and air conditioning equipment with inverter technology was acquired, and 100% of the cold rooms at the Meliá Varadero Hotel were sealed.

Within the most relevant lessons learned it could be remarked: The unprecedented socialization that allowed unity of criteria and exchange of experiences; energy companies working with management systems; and the dynamism and professionalism of the parties in the workshops and activities carried out. In addition, some of the participants highlighted how useful was the elaboration of baselines, learning about the ISO 50001:2018, and the role of measuring energy parameters. They also remarked to have learned that "success is continuous improvement, writing what you do and doing what you write."

In the case of Ecuador, the project was promoted by the Ministry of Energy and Mines, supported by the EUROCLIMA+ Program through the GIZ, and aimed to promote energy management within the industrial sector as a strategy to improve their contributions to the commitments acquired in the NDC. In 2021, contact with companies in the industrial sector was initiated to identify their interest in participating in the LEEN. By mid-2022, the Ministry made the call, and 16 industrial companies with high and medium energy consumption were registered to the project.

In July 2022, the first LEENs workshop was held. During the workshop, commitment documents were signed by industry representatives and ministerial authorities. As part of the process, eight workshops and four webinars designed to address the specific needs of the network were conducted. Additionally, they had meetings with the senior management of the companies to increase commitment. In some cases, they achieved to obtain a certification in energy management.

One of the most relevant goals of the Ecuadorian case was the high level of progress in the implementation of the EMS: Six companies expected to reach between 75% and 80%; five companies expected to reach between 91% and 95%; three companies expected to reach between 96% and 100%; and one company did not set a concrete goal for advancement. To reach 100% implementation of their EMS, six companies set a deadline of October 2023: three in December 2023, two in March 2024, and four by June 2024. Four of the 15 companies established a target to certify their EMS with ISO 50001:2018 by June 2023 and the rest between November 2023 and March 2024. A total of 77 energy performance improvement measures were proposed, of which 31 (40%) were implemented before the end of the G&A work in July 2023, 28 (36%) were implemented by the end of 2023, and the remaining 18 (24%). They will be fully implemented by the end of 2024.

As mentioned by the professional interviewed for the Nicaragua chapter, *"LEENs in Nicaragua have helped to understand that in the world we are all connected trying to solve problems... what we do not know is that others have already done it or at least have tried and that it is easier to continue what they have done than to start from scratch. But do not realize by countless factors, forgetting that we are all connected and that several heads think better than one."* Even though it is a novel experience, LEENs seem to have started to break down many barriers in Nicaragua, mainly related to the lack of confidence in sharing information. However, one of the key challenges confronted was related to the share of information among participants.

7.2 Learning Energy Efficiency Networks and Energy Management System as a Promising Tool for Energy Efficiency

Worldwide, the EEN and/or the LEENs have been demonstrated to be effective instruments for promoting the implementation of energy efficiency measures because of their quantitative and qualitative impact. The companies that participate in the Learning Networks implement EMS in less time and at lower costs than those that do it individually, generate valuable links that help to fulfill energy improvement objectives, and develop the individual capabilities of their employees and promote cultural change as a result of the exchange of information, experiences, and knowledge, and implementation of an EMS.

The regional cases described in this book show that, in addition to energy savings, the most important results of LEENs in developing countries are, indeed, ***changes in cultural patterns, increasing efficiency awareness, and capacity building***. In this sense, networks have played a fundamental role in removing the main barriers to energy efficiency in regional industrial and commercial sectors.

The results obtained have been very satisfactory for the companies and professionals who have attended all the learning workshops. Professionals have worked on different energy analysis tools that will allow them to grow in their respective jobs, achieving added motivation to their usual responsibilities. However, the great contribution that the LEENs offer is the cultural change among the professionals.

The experiences documented in this book may serve as some lessons learned and good practices to be replicated in future networks in the region, and/or in other developing regions.

First, it is crucial to design a communication strategy to both guarantee the long-term participation of professionals in the LEEN and to promote collaborative work, helping companies to understand that working together is more productive than trying to solve everything by themselves. This collaborative work also helps horizontal knowledge transmission. The communication strategy must be designed from the very beginning. Most LAC countries have developed a two-sided strategy, including informal and formal communication channels. Informal communication (usually deployed by chat groups) helps to increase the confidence between participants and share horizontal information and capacity buildings. In general, chat groups include the organizer, the sponsor, and the companies, and demonstrated to be very useful in making quick and effective decisions in the face of adversities that arise during the networks. In the case of formal communication, it includes calls, agendas, and so on. All the information must be available to participants on time, so they have enough capacity to go through it and be ready to participate and review all the work done if necessary. This is usually performed by the coordinator and moderator of the LEEN.

Second, the network helps participants realize that, usually, challenges are similar between different companies, regardless of size. This gave them more confidence to move forward together and collaborate to learn from companies ahead and learn about their experience in the processes. Many countries have highlighted that the exchange of experiences of companies that have carried out innovative projects and implemented actions in their facilities, which allowed them to generate procedures and tools that are especially useful to boost their EMS during workshops, were very effective in helping in the learning process of other companies to understand of the needs of the management system.

Third, the motivation of each participant and company's commitment are key to the success of the LEEN. Only LEENs that achieved long-term participation of professionals along the whole process have shown good results. Indeed, in some cases, one of the most relevant barriers confronted by the networks has been to maintain the participation of companies and professionals along the process. In this regard, it is important to have different spaces to regularly share the progress of companies in the process. This is useful to encourage the most backward companies to catch up with pending tasks and to allow corrective actions to ensure the expected results of the network. In some cases, the participation of both the initiator, and the sponsor or financing agency, in some workshops, in addition to the energy consultant in the face-to-face meetings held throughout the work of the network has been very effective in giving institutional presence and achieving greater commitment in the participation of companies in the program.

Fourth, workshops and visits to companies are crucial in this methodology, and therefore they must be well-planned and developed. Indeed, these meetings are key to promoting the collaborative work of the LEENs, inspiring trust between participants, which is crucial to enable exchanging information and experiences among participants as remarked in previous paragraphs. This aspect has been very important in some of the cases to remove the barrier of lack of trust. Workshops must be, therefore, well designed in advance, including the technical topics to be taught in the different meetings, with the presence of the energy experts and all the participants.

Fifth, the coordinator, moderator, and technical support/ energy consultant must act as a consolidated group throughout the process. Each one of these profiles is key to the success of the network. This group defines and coordinates the activities throughout the network and must consider the needs of the participants to ensure excellent results in the implementation of the program. The moderator role is key in the workshops and the communication strategy, to get the commitment and confidence of participants. Technical support is crucial so that the companies can talk about the technical issues raised in the sessions and generate group instances to clarify doubts and share experiences.

Sixth, energy diagnostics are fundamental in the process, as they are the basis for building the base line and identifying opportunities and measures. Then it is advisable to carry out energy diagnostics as early as possible at the start of the network and to have a homogeneous methodology to perform them by local technical assistants. Some experiences show that involving local technical assistants in energy diagnostics help build domestic technical capabilities to serve the country's consulting market, and promote energy efficiency in the business sector. This aspect is very important in the case of developing regions, which are initiating an energy efficiency market. In a similar direction, involving companies in the identification of measures from an early stage seen as a good practice and helps in the establishment of action plans.

Finally, monitoring the progress of the LEEN along the process is very important to evaluate results and correct any situation that needs to be modified. It is important to monitor and evaluate progress against goals throughout the network. A permanent monitoring process allows timely adjustments to achieve the goals set at the beginning of the LEEN.

This work contains a set of lessons learned from the experience of the LAC that have implemented these networks that may be useful for other countries of the region and countries in other developing regions. Barriers to the development and implementation of energy efficiency that LEEN can mitigate or eliminate are very similar for these countries, due to their socio-economic, financial, and environmental characteristics.

However, it is important to highlight that the performance of the LEENs, and many other measures and strategies depend on the context of implementation, boundary conditions, and enabling conditions. This is why it is crucial to adapt the LEEN to the local situation. The self-design of the LEENs should be actively promoted as part of the development of public policies to improve energy management, given the significant benefits and auxiliary benefits mentioned.

LEENs can be a very attractive tool for governments to knock down some of the most common barriers in industrial, commercial, and services sectors, and they have shown to have positive impacts in the social and economic performance of the companies.

References

Allcott, H., Greenstone, M. (2012). "Is There an Energy Efficiency Gap?" *Journal of Economic Perspectives*, 26 (1), pp. 3–28.

Azpiazu, D. (2008). "Concentración y centralización del capital en el mercado eléctrico Argentino," *Realidad Económica*, 233, pp. 86–110.

Basualdo, E., Daniel, A. (2002). "El proceso de privatización en la Argentina. La renegociación con las empresas privatizadas. Revisión contractual y supresión de privilegios y rentas extraordinarias." *Área de Economía y Tecnología de la FLACSO*, Sede Argentina, pp. 42.

Bhatia, M., Angelou, N. (2015). "Beyond Connections: Energy Access Redefined."*Technical Report 008/15*. Energy Sector Management Assistance Program / The World Bank.

Bouille, D. (1999). "Lineamientos para la Regulación del Uso Eficiente de la Energía en Argentina." Serie Medio Ambiente y Desarrollo CEPAL N° 16, Proyecto CEPAL/COMISIÓN EUROPEA "Promoción del Uso Eficiente de la Energía en América Latina," Santiago de Chile, marzo 1999.

Bouille, D., Behnisch, A., Bravo, G., Di Sbroiavacca, N., Dubrovsky, H., Lallana, F., Nadal, G., Pistonesi, H., Recalde, M., Romano, A., Ruchansky, B., Sagardoy, I., Landaveri, R.(2021). "INFORME FINAL DE LA PROPUESTA DEL PLAN NACIONAL DE EFICIENCIA ENERGÉTICA ARGENTINA (PlaNEEAr)," Proyecto de Eficiencia Energética Argentina. GFA Consulting Group.

Bourges, C. (2013). "Política Nacional de Eficiencia Energética. Breve Resumen de las últimas tres décadas 1980–2010". Instituto Argentino de la Energía General Mosconi y Consejo Profesional de Ingeniería Mecánica y Electricista, Buenos Aires, agosto 2013.

Brown, M. (2014). "Innovative energy-efficiency policies: An international review WIREs Energy Environ 2014."

Calleja, G. A. (2005). "Política energética del gobierno Alfonsín," Realidad económica 214, Buenos Aires, pp. 105–128.

Calleja, Gustavo. A. (2005)."Política energética del gobiernos Alfonsín," *Realidad económica*, 214, pp. 105–128.

Camarda, M. F. (2019). "La Institucionalidad de la Eficiencia Energética en la República Argentina: Un Análisis de las Principales Políticas en el período 1980–2017". Revista Administración Pública y Sociedad, Instituto de Investigación y Formación en Administración Pública (IIFAP-FCS-UNC), N° 7, *enero-junio 2019*, pp. 22–43.

Carpio, C., Recalde, M. Y. (2020). "Learning energy efficiency networks in Latin America: Lessons learned from the Argentinean case," *WIREs—Wiley Interdisciplinary Reviews: Energy and Environment.*

Carpio, C., Recalde, M. Y. (2021). "Learning energy efficiency networks in Argentina," *Advanced Science Reviews.*

Chassein, E., Durand, A., Gerspacher, A., Jochem, E., Roser, A.(2018). "Evaluation of regional Learning Energy Efficiency Networks—Generation, Audits, Targeting, and Network Operation. Working Paper No. 2," Karlsruhe, Germany: IREES.

CONUEE. (2019). "PROGRAMA NACIONAL PARA SISTEMAS DE GESTIÓN DE LA ENERGÍA 2013–2018."https://www.conuee.gob.mx/transparencia/boletines/INFORMES/SGEs/Informe_PRONASGEn_Final_30042019.pdf

De Buen et al.(2003)."A strategy for energy efficiency actions in the Mexican industrial sector: The Pemex experience. "https://policycommons.net/artifacts/2230818/a-strategy-for-energy-efficiency-actions-in-the-mexican-industrial-sector/2988250/

Dubrovsky, H., Recalde, M. Y., Bouille, D. H., Nadal, G., Bravo, G. Behnish, A. (2020). "Principales ramas de la industria manufacturera desde la perspectiva de la eficiencia energética," *Proyecto de Eficiencia Energética Argentina*. GFA Consulting Group.

Duguma, L. A., Wambugu, S. W., Minang, P. A., Noordwijk, M. (2014). "A systematic analysis of enabling condition for synergy between climate change mitigation and adaptation measures in developing countries," *Environ. Sci. Policy* 42, pp. 138–148.

Duzgun, B., Komurgoz, G. (2014). "Turkey's energy efficiency assessment: White Certificates Systems and their applicability in Turkey," *Energy Pol.* 65, pp. 465–474.

Dyner, I., Olaya, Y., Franco, C. J.(2011). "An enabling framework for wind power in Colombia: What are the lessons from Latin America?"

Diffusion of Renewable Energy Technologies: Case Studies of Enabling Frameworks in Developing Countries, pp. 73–86.

Ellis, J. (2010). *Energy Service Companies (ESCOs) in Developing Countries.* International Institute for Sustainable Development, Manitoba, Canada.

Gerarden, T. D., Newell, R. G., Stavins, R. N. (2017). "Assessing the Energy-Efficiency Gap," *Journal of Economic Literature*, 55 (4), pp. 1486–1525.

Gillingham, K., Palmer, K. (2014). "Bridging the Energy Efficiency Gap: Policy Insights from Economic Theory and Empirical Evidence," *Review of Environmental Economics and Policy*, 8(1), pp. 18–38.

Giraudet, L. G., Missemer, A. (2023)."The History of Energy Efficiency in Economics: Break Points and Regularities," *Energy Research & Social Science*, pp. 102973.

GIZ. (2016). "DOCUMENTO MEMORIA DE LA RED DE APRENDIZAJE PARA LA IMPLEMENTACIÓN DE SISTEMAS DE GESTIÓN DE LA ENERGÍA EN LA INDUSTRIA".https://www.giz.de/en/downloads/giz2016-es-Documento_Memoria_de_la_Red_de_Aprendizaje_SGEn_Industria.pdf

Grubb, M., McDowall, W., Drummond, P. (2017). "On order and complexity in innovations systems: Conceptual frameworks for policy mixes in sustainability transitions," *Energy Research & Social Science*, 33 (2017), pp. 21–34.

Guzowski, C., Recalde, M. (2008). "Renewable energy in Argentina: Energy policy analysis and perspectives," *International Journal of Hydrogen Energy*, 33 (13), pp. 3592–3595. (Julio, 2008).

Guzowski, C., Recalde, M. (2008b). "Barreras a la entrada de las Energías Renovables: el caso Argentino." *Avances en Energías Renovables y Medio Ambiente (AVERMA)*, 12, pp. 12.31–12.38. Asociación Argentina de Energías Renovables y Ambiente.

Helming, S., Göbel, M. (1998). *Planificación de proyectos orientada a objetivos ZOPP*, Eschborn: Multiprint Frankfurt, pp. 1–35.

IEA. (2023). Energy Efficiency 2023, IEA, Paris. https://www.iea.org/reports/energy-efficiency-2023, License: CC BY 4.0.

IPCC.(2022)."Summary for Policymakers," In: *Climate Change 2022: Mitigation of Climate Change.* Contribution of Working Group III to the Sixth Assessment Report of the Intergovernmental Panel on Climate Change [P. R. Shukla, J. Skea, R. Slade, A. Al Khourdajie, R. van Diemen, D. McCollum, M. Pathak, S. Some, P. Vyas, R. Fradera, M. Belkacemi, A. Hasija, G. Lisboa, S. Luz, J. Malley, (eds.)]. Cambridge University Press, Cambridge, UK and NY, USA.

Jacobsson, S., Bergek, A., Sandén, B. (2017). "Improving the European Commission's analytical base for designing instrument mixes in the energy sector: Market failures versus system weaknesses," *Energy Research & Social Science*, 33 (2017), pp. 11–20.

Jaffe, A. B., Stavins, R. N. (1994). "The energy-efficiency gap: What does it mean?"*Energy Policy*, 22 (10), pp. 804–810.

Jimenez, R., Yépez-García, A. (2016). "Composition and sensitivity of residential energy consumption (No. IDB-WP-690)," IDB Working Paper Series.

Jimenez, R., Yepez-Garcia, A. (2017). "Understanding the drivers of household energy spending: Micro evidence for Latin America (No. IDB-WP-805)," IDB Working Paper Series.

Jochem, E., Gruber, E. (2007). "Local learning-networks on energy efficiency in industry—Successful initiative in Germany," *Applied Energy*, 84 (2007), pp. 806–816.

Kern, F., Kivimaa, P., Martiskainen, M. (2017). "Policy packaging or policy patching? The development of complex energy efficiency policy mixes," *Energy Research & Social Science*, 23, pp. 11.

Köwener, D., Jochem, E., Mielicke, U. (2011). "Energy Efficiency Networks for companies—Concept, achievements, and prospects," *ECEEE Proceedings*, 2011(2), pp. 725–733.

Köwener, D., Nabitz, L., Mielicke, U., Idrissova, F., Fraunhofer Institut für System- und Innovations forschung (2014). "Learning energy efficiency networks for companies—saving potentials, realization and dissemination," In: *ECEEE 2014 Industrial Summer Study on Energy Efficiency. Proceedings: Retool for a Competitive and Sustainable Industry*. Stockholm, pp. 91–100.

Kozulj, R. (2002). "Balance de la privatización de la industria petrolera en Argentina y su impacto sobre las inversiones y la competencia en los mercados minoristas de combustibles," *CEPAL, Serie Recursos Naturales e Infraestructura*, 46, pp. 1–84.

Langlois-Bertrand, S., Benhaddadi, M., Jegen, M., Pineau, P. O. (2015). "Political-institutional barriers to energy efficiency," *Energy Strategy Reviews*, 8, pp. 30–38.

Liu, X., Shen, B, Price, L., Hasanbeigi, A., Lu, H., Yu, C., Fu, G. (2019)."A review of international practices for energy efficiency and carbon emissions reduction and lessons learned for China," *WIREs Energy Environ.*, 2019, pp. e342.

Nygaard, I., Hansen, U. E. (2015). *Overcoming Barriers to the Transfer and Diffusion of Climate Technologies*, 2nd ed., UNEP DTU Partnership.

TNA Guidebook Series. https://orbit.dtu.dk/ws/files/121688225/Overcoming_Barriers_2nd_ed.pdf

Nygaard, I., Hansen, U. E. (2015). "The conceptual and practical challenge to technology categorization in the preparation of technology needs assessments," *Clim. Change*, 131 (3), pp. 371–385.

Ockwell, D., Sagar, A., de Coninck, H. (2015)."Collaborative research and development (R&D) for climate technology transfer and uptake in developing countries: Towards a needs driven approach," *Clim. Change*, 131 (3), pp. 401–415.

OECD / IPEEC.(2017). "Energy Efficiency Networks: Towards good practices and guidelines for effective policies to stimulate energy efficiency," Working Paper.

Pistonesi, H. (2000). "Sistema eléctrico argentino: los principales problemas regulatorios y el desempeño posterior a la reforma," CEPAL. Santiago de Chile: Naciones Unidas.

Pistonesi, H. (2001). "Desempeño de las industrias de electricidad y gas natural después de las reformas: el caso de Argentina," CEPAL. Santiago de Chile: Naciones Unidas.

Purkus, A., Gawel, E., Thrän, D. (2017). "Addressing uncertainty in decarbonisation policy mixes—Lessons learned from German and European bioenergy policy," *Energy Research & Social Science*, 33 (2017), pp. 82–94. https://doi.org/10.1016/j.erss.2017.09.020.

Recalde, M. (2011). "SISTEMAS ENERGÉTICOS, MERCADO Y ESTADO: El rol de los recursos naturales energéticos y la política energética en el caso argentino," Editorial Académica Española, Madrid, España.

Recalde, M. (2011)."Determinantes de la inversion en exploración de hidrocarburos: Un análisis del caso argentino," *Cuadernos de Economía—Spanish Journal of Economics and Finance*, 34 (94), enero-abril 2011, pp.40–52. ELSEVIER DOYMA (ISSN: 0210-0266).

Recalde, M. (2011)."Energy policy and energy market performance: The Argentinean case," *Energy Policy*, 39 (6), pp. 3860–3868.PERGAMON-ELSEVIER SCIENCE LTD (ISSN: 0301–4215).

Recalde, M. (2017a). "La transición energética hacia las energías renovables en América Latina," capítulo del libro: "Ambiente, agua y energía: Aportes jurídicos para su relación—1a ed". Ed. Lajouane, pp. 195–244.

Recalde, M., Ramos-Martin, J. (2012). "Going beyond energy intensity to understand the energy metabolism of nations: the case of Argentina," *Energy*, 37 (1), pp. 122–132. PERGAMON-ELSEVIER SCIENCE LTD (ISSN: 0360-5442).

Recalde, M. Y., Bouille, D., Girardin, L. O. (2015). "Limitaciones para el desarrollo de energías Renovables en Argentina: el rol de las condiciones de marco desde una perspective histórica," *Revista Problemas del Desarrollo*, 183 (46), octubre-diciembre 2015, pp. 89–115.

Recalde, M. Y., Zabaloy, F., Guzowski, C. (2018). "El Rol de la Eficiencia Energética en el Sector Residencial para la Transición Energética en la Región Latinoamericana," TRAYECTORIAS AÑO 20, NÚM. 47 | JUL-DIC 18, pp. 77–102.

Recalde, M., Guzowski, C. (2012). "Boundaries in promoting energy efficiency: Lessons from the Argentinean case," *International Journal of Hydrogen Energy*, 37(19), pp. 14725–14729.

Recalde, M. Y. (2017). "La inversión en energías Renovables en Argentina," *Revista de Economía Institucional,* 19 (36), pp. 231–254.

Recalde, M. Y., Ramos-M, J.(2012). "Going beyond energy intensity to understand the energy metabolism of nations: The case of Argentina," *Energy*, 37 (1), pp. 122–132.

Reister, D.(1987). "The link between energy and GDP in developing countries," *Energy*, 12 (6), pp. 427–433.

Rohde, C., Mielicke, U., Nabitz, L. (2015)."Learning Energy Efficiency Networks—Evidence based experiences from Germany," *ACEEE Summer Study on Energy Efficiency in Industry*.

Rosenow, J., Eyre, N. (2022). "Reinventing energy efficiency for net zero," *Energy Research & Social Science*, 90.

Rosenow, J., Fawcett, T., Eyre, N., Oikonomou, V. (2016). "Energy efficiency and the policy mix," *Building Research & Information*, 44 (5–6), pp. 562–574.

Rosenow, J., Kern, F., Rogge, K. (2017). "The need for comprehensive and well targeted instrument mixes to stimulate energy transitions: The case of energy efficiency policy," *Energy Research & Social Science*, 33, November 2017, pp. 95–104.

Ruchansky, B., Januzzi, G., Buen, O. D., Romero, A. (2011). "Eficacia institucional de los programas nacionales de eficiencia energética: los casos del Brasil, Chile, México y el Uruguay," CEPAL, Santiago de Chile. Retrieved from: https://repositorio.cepal.org/bitstream/handle/11362/6355/2/S1100313_es.pdf

Safarzadeh, S., Rasti-Barzoki, M., Reza Hejazi, S. (2020)."A review of optimal energy policy instruments on industrial energy efficiency programs, rebound effects, and government policies, "*Energy Policy*, 139, 2020, pp. 111342.https://doi.org/10.1016/j.enpol.2020.111342.

Schlomann, B. (2016). "Energy Efficiency Networks," Policy Brief. ODYSSEE MURE project. https://www.odyssee-mure.eu/publications/archives/

Soren, A., Newell, R. (2004). "Information programs for technology adoption: The case of energy-efficiency audits," *Resource and Energy Economics*, 26 (1), pp. 27–50.

Sorrell, S. (2004). *The Economics of Energy Efficiency: Barriers to Cost-Effective Investment*. Cheltenham: Edward Elgar.

Sorrell, S., Mallett, A., Nye S. (2011). "Barriers to industrial energy efficiency: A literature review," DEVELOPMENT POLICY, STATISTICS AND RESEARCH BRANCH WORKING PAPER 10/2011.

Stadelmann, M. (2017). "Mind the gap? Critically reviewing the energy efficiency gap with empirical evidence," *Energy Res. Soc. Sci.*, 27, pp. 117–128.

Stede, J. (2017). "Bridging the industrial energy efficiency gap—assessing the evidence from the Italian white certificate scheme," *Energy Pol.*, 104, pp. 112–123.

Stede, J. (2019). "Do energy efficiency networks save energy? Evidence from German plant-level data," DIW Discussion Papers, No. 1813, Deutsches Institut für Wirtschaftsforschung (DIW), Berlin.

Tahvonen, O., Salo, S. (2001). "Economic growth and transitions between renewable and non-renewable energy resources," *Eur. Econ. Rev.*, 45 (8), pp. 1379–1398.

Tanaka, K. (2011). "Review of policies and measures for energy efficiency in industry sector," *Energy Policy*, 39(10), pp. 6532–6550. doi:10.1016/j.enpol.2011.07.058.

Thollander, P., Backlund, S., Trianni, A., Cagno, E. (2013). "Beyond barriers—A case study on driving forces for improved energy efficiency in the foundry industries in Finland, France, Germany, Italy, Poland, Spain, and Sweden," *Appl. Energy*, 111, pp. 636–643.

Trærup, S., Stephan, J. (2015). "Technologies for adaptation to climate change. Examples from the agricultural and water sectors in Lebanon," *Clim. Change,* 131 (3), pp. 435–449.

Trianni, A., Cagno, E. (2012). "Dealing with barriers to energy efficiency and SMEs: some empirical evidences," *Energy*, 37, pp. 494–504.

Trianni, A., Cagno, E., Farné S. (2016). "Barriers, drivers and decision-making process for industrial energy efficiency: A broad study among manufacturing small and medium-sized enterprises," *Applied Energy*, 162, pp. 1537–1551.

Villegas Alcántar, N., Jáuregui Nares, I., de Buen Rodriguez, O., Lameiras Barrera, D., Atala Palacios, J. E., Skwierinski Durán, A. M., Feilbogen, E. (Ed). (2016). "Documento Memoria de la Red de Aprendizaje para la implementación de Sistemas de Gestión de la Energía en la Industria México," CDMX., septiembre del 2016. Conuee / GIZ. https://www.giz.de/en/downloads/giz2016-es-Documento_Memoria_de_la_Red_de_Aprendizaje_SGEn_Industria.pdf

Vogel, J. A. Per Lundqvist, P., Arias, J. (2015). "Categorizing barriers to energy efficiency in buildings," *Energy Procedia*, 75, pp. 2839–2845.

Watson, J. Byrne, R., Ockwell, D., Stua, M. (2015). "Lessons from China: Building technological capabilities for low carbon technology transfer and development," *Clim. Change,* 131 (3), pp. 387–399.

Yeatts, D., Auden, D., Cooksey, C., Chen, C. F. (2017). "A systematic review of strategies for overcoming the barriers to energy-efficient technologies in buildings," *Energy Res. Soc. Sci.* 32, pp. 76–85.

Zabaloy, F., Recalde, M., Guzowski, C. (2019). "Are energy efficiency policies for household context dependent? A comparative study of Brazil, Chile, Colombia and Uruguay," *Energy Research & Social Science (ERSS)*, 52 (2019), pp. 41–54.

Zhu, J., Chertow, M. R. (2017). "Business strategy under institutional constraints: Evidence from China's energy efficiency regulations," *Ecol. Econ.,* 135, pp. 10–21.

Index

For Product Safety Concerns and Information please contact our
EU representative GPSR@taylorandfrancis.com Taylor & Francis
Verlag GmbH, Kaufingerstraße 24, 80331 München, Germany